THE NUN IN THE WORLD

THE NUN IN THE WORLD

New Dimensions in the Modern Apostolate

by

LEON JOSEPH CARDINAL SUENENS

Archbishop of Malines-Brussels

THE NEWMAN PRESS
WESTMINSTER·MARYLAND

Translated from the French by GEOFFREY STEVENS

FOREWORD

THE Council is an invitation to an examination of conscience. It is a matter of getting clear what is essential in the Church and what is only incidental, what must remain and what is dependent on the times and circumstances. This sorting out is necessary in many sections of the Church's pastoral work.

With this idea in mind we hope to examine in these pages the place and the mission of nuns in the Church in terms of the world as it is today. This study is of interest to the whole Church, for the nun is called to play a leading role. In putting the emphasis on the part she has to play, we shall none the less bring to light the chief problems that concern the whole field of pastoral work.

The study should also give us a better understanding of the meaning of religious vocation by freeing it of the anachronisms which fetter it at present.

We shall be concerned only with nuns belonging to orders and congregations dedicated to apostolic work. The contemplative life, which is on a different plane altogether, is not considered here.

When we talk of nuns we have also in mind all souls dedicated to God whose object is visible apostolic propagation and who, though their methods and forms are different, share the same basic vocation. We have also made free use of the word 'religious'—albeit with feminine pronouns—to indicate that, *mutatis mutandis*, what is said applies also to male religious who are not priests.

It is of all these chosen souls that we are thinking in this book. May the Council mark for them too the dawn of the spiritual spring-time of which Pope John XXIII has spoken. In the analysis of their present position—to which we bring as much love and respect as we do candour—our aim is to help them deploy on a global scale the inexhaustible treasure of their devotion and the wealth of their spirituality.

A word about the title. We speak about 'the Nun in the World' and 'new dimensions in the modern apostolate'. Let it be quite clear that it is not the apostolic role of the nun that is new in itself. Far from it. Religious are apostolic by dedication when they take their vows. But the modern world calls for a modern approach, new dimensions, and it is the integration of nuns into these new dimensions which concerns us.

In the same way when we speak of the 'apostolic advancement' of nuns, we do not mean an advancement from a non-apostolic to an apostolic role, but a re-deployment of their talents in the new dimensions. This, incidentally, does not mean that we are suggesting a heavier load for their already overcrowded time-tables: we are pleading rather for a pruning of them and proper attention to the scale of values. A proper balance must be obtained between the essential and indispensable components of the religious life and the priorities of what God expects of us. One of these components is indeed the apostolic aspect which this book tries to bring out. We shall outline a sort of theology of the day to apply to the life of nuns, and we shall do so with our minds firmly fixed on God's requirements.

Malines,
15th September, 1962
Feast of the Seven Dolours

CONTENTS

Part I: THE POSITION TODAY

Part II: TOWARDS AN APOSTOLIC RENEWAL OF THE RELIGIOUS LIFE

LETTER FROM THE VATICAN
SECRETARY OF STATE

Secretario di Stato
di Sua Santità

Vatican
29 November, 1962.

Your Eminence,

The Sovereign Pontiff has learnt that Your Eminence, having already shed light on various aspects of the Apostolate of the Church in a series of extremely pertinent works, has now been working on a study of the role of the nun in the world today.

Well aware of the pastoral zeal by which Your Eminence is inspired, His Holiness is greatly pleased to see Your Eminence devoting his time and talents to a problem which today forms the main concern of many pastors. There can, indeed, be no doubt that the present conditions of the Apostolate indicate the need for useful reforms in certain spheres, particularly a better utilization of all the apostolic forces at the Church's command.

Among these, nuns hold a place of honour and every encouragement should be given to any effort which will enable the spiritual treasures they possess to be placed more directly at the service of the Apostolate.

His Holiness takes pleasure in thinking that the study prepared by Your Eminence will shed new and valuable light on this question and as token of his fatherly goodwill He sends you with this letter his wholehearted and paternal apostolic blessing.

Please accept, Your Eminence, the expression of my highest esteem with which, venerating the sacred purple,

I have the honour to be once more,

Your Eminence's most humble, devoted and obedient servant in Jesus Christ,

✠ A. Dell' Acqua,
Subst.

Part I

THE POSITION TODAY

1

THE WORLD WE LIVE IN

THE TEMPO OF THE WORLD

In order that the position of women in religion may be accurately assessed in the contemporary world, it is necessary to consider both them and the world in which they have to live. These two distinct realities have their own separate internal evolutionary laws, their own vital rhythm. Religious in the world cannot occupy a valid place without taking into account the evolution of modern society. Yeast is not placed beside the dough it is to leaven, but right in it. Somehow, then, it is necessary that the two vital rhythms be brought into some harmonious relationship.

Contemporary society, our society, the one in which we live and which it is up to us to save, which Christ has entrusted to us, is a society evolving with bewildering rapidity in all its aspects at once. It is characterized by a set of sociological phenomena which we must recognize if we are not to remain on the fringe of things and miss all dialogue with the world for lack of contact or even a common language.

The world of today differs from that of yesterday in a number of ways. We shall consider at length only those differences which have a more direct effect on women in religion.

A CHANGING WORLD

Our world is a world in a continuous state of becoming,

3

a state in which everything is questioned. We no longer live in an age where daily lives were solidly framed in tradition, and institutions were there to safeguard values that were never called in question. On all sides tradition is foundering and conformism falling in ruins. The world of today wants to re-think and revise everything, to take the universe apart to see what makes it tick. The emphasis is on personal values, on freedom, responsibility, invention and creation.

The man of today has seen everything, read everything, heard everything; he is determined to explore everything, greedy of experience, anxious to set foot in unexplored fields.

THE WORLD AS A UNIT

Our planet has seen more revolutionary development in the last quarter of a century than in all the preceding nineteen. In the technical sphere, from the first atom-bomb to the interplanetary rockets of today, man has made giant strides and these are probably no more than a beginning, a curtain-raiser for what is to come.

In this world in which geographical distances have ceased to have any meaning, we are experiencing the hatching of new concepts and new ideologies. We hear a televised broadcast of a U.N. session before those present in person, since radio waves travel quicker than sound waves. Anyone may see with his own eyes and hear with his own ears international conferences taking place in Tokyo or Melbourne. Every evening we can tour the world sitting in an armchair before our TV set. No aspect of human life remains alien to us. Through all the mass-communication media—press, radio, cinema—we are subjected to psychological and moral influences attempting to mould the man of today to the image of this developing society. Matters are not treated on a provincial scale nowadays, nor even

on a national or continental scale, but on an inter-
continental scale—until such time as it is superseded by
the interplanetary scale. To set a problem in other than
global terms, said a certain statesman, is to set it on un-
sound premisses.

And that, whether we like it or not, is how we stand
today.

A WORLD ON THE MOVE

In this world the human ant-heap is moving in all direc-
tions. Flying from Europe to America is a matter of a few
hours. One has only to spend a few moments at an air
terminal to realize the extent to which the world has
become a cross-roads of peoples. On top of this, we are
experiencing considerable internal migration and the rural
world is more and more giving place to an urban civilization. Millions of uprooted country people swarm to the
suburbs of our octopus cities. This move alone loosens their
religious roots, without mentioning the moral repercussions
of crowding masses of human beings into inadequate space
with all the disastrous consequences to family life which
this entails.

NEW VALUES

It is not only in the field of technical progress that our
world has become a 'new' world in comparison with that
of twenty, thirty or forty years ago; men in our times have
acquired new standards. The old values have been devalued
and new ones have taken their place.

Men aspire to equality among themselves, to fraternity
and solidarity. Feudalism is a thing of the past and Queen
Anne is really dead. Certain social distinctions are no longer
accepted and privilege is denied. Everything which reminds
people of the manners of an earlier age, the artificial
etiquette of Courts, certain bourgeois customs and conven-

tions, marks of respect in so far as they are pompous or obsequious—all these belong now to the past.

Men look for candour, efficiency, directness, sincerity, brevity and sobriety. Art must be deprived of its flourishes and useless complexities. Human relationships are being rediscovered.

AN AMBIGUOUS WORLD

We must be able to understand this evolution and to distinguish what it holds of truth if we are to have any success in the task which we must undertake of sorting it out and judging it objectively in the spirit of faith.

The 'world', whose atmosphere threatens to permeate us if we are content to suffer it instead of trying to guide it, is criss-crossed by influences which are as powerful as they are to be feared. Technical progress is one thing; the use to which mankind puts it is something else again. The day after Hiroshima, Denis de Rougemont wrote, "The bomb is not dangerous; it is a thing . . . ; what is terrifyingly dangerous is man".

The astounding success of technical achievement intoxicates our contemporaries who scarcely fall short of idolatry in their respect for science. Gagarin, the man in the first sputnik, was told by his masters to declare that the heavens were empty, that he had not met God in orbit. Men today can hardly bring themselves to admit the existence of anything they cannot calculate or measure. They can no longer see God where He is, infinitely close to them, in the depths of their innermost being, giving them life, sense and purpose. They look for Him through a telescope, if indeed they seek Him at all, so closed are their minds to what they cannot see, to mysteries beyond their comprehension.

Mankind, thus deprived of its roots, is dragged helplessly by the fearful tides of materialism.

This materialism, whether it be admitted or remain under

cover, does its best to rivet man to earth, to deprive him of his soul, and to block his spiritual horizon. These are evil times for the disinterested vocations where money is not the supreme criterion of success. Everything conspires to rob man of his sense of perspective and proportion. "What does it profit a man if he gain the whole world and suffer the loss of his own soul?" These words no longer have any meaning for the masses who never raise their eyes to Heaven. Material well-being absorbs all their energies and becomes an end in itself.

Man suddenly sees his standard of living raised—in our countries, at least—and he begins to have more free time at his disposal. Work he accepts, but he expects also relaxation, week-ends, paid holidays and so forth; a continual increase in leisure is an integral part of existence. The nervous tension which is a feature of modern life demands holidays in the open air, sport and physical culture. Every evening people switch on the TV and allow entry to whatever the waves bring, be it healthy or unhealthy, as if they were a beach open to whatever the waves of the sea might wash up.

What is man going to do with this leisure, which he hails as human and social progress? What chance is there for keeping the Christian Sunday in the mass week-end exodus from the towns? Books, films and broadcasts which should be enriching our culture, what message have they? What answer do they propose to the fundamental questions man asks himself about his own nature, about the problems of suffering and death and the beyond? Do these mass-communication media not in fact merely pour forth wave upon wave of smutty 'literature' extolling free love, irreligion, divorce, euthanasia, relativism and indifferentism?

Our age is full of positive values, but also full of ambiguities. We must love it, yet defend ourselves against it; collaborate with its enthusiasm, yet canalize it; admire it, yet set limits to it; encourage it along every path that can lead to good, yet warn it of the gulfs and precipices that flank the way. We must distract man from himself and make him aware of the Saviour who remains for him, too, in the heart of the twentieth century, 'the Way, the Truth and the Life'.

The Christian of today must exercise a vital and clear-sighted Christianity which will demand of him certain alliances and certain reservations, certain adherences and certain denials, a generosity of views that will enable him to welcome all manifestations of brotherly goodwill, and a charity secure from compromise which will move him to share with others the treasures of life which he enjoys.

Every great speaker, it has been said, has two geniuses, his own and that of his times. The same is true of every Christian vowed at his baptism to the work of the aposto-late. He must have his own genius—that is, he needs the Spirit of the Lord to enlighten him and feed his enthusiasm —but he must also know the genius of his time, its suscep-tibilities and its reflexes. Since he belongs to two worlds, that of God and that of mankind, he can only be the mediator between them if he is at the same time loyal to both.

The Christian life is a paradox at any time. But it re-quires a maturity, a clear-sightedness and an increased sense of responsibility in the world in which we live today and which it is our job to christianize.

It was necessary to do this rapid sketch to serve as back-ground to the following chapters. This is the context in which women in religion have to play their part and find their place. But before going on to a consideration of their role in the world, we must say what has happened to woman in the world of today, what her place is in the current evolutionary process. A religious is a woman of her times, so the question of woman's status is a vital one for her. She cannot renounce her femininity, and it is in fact this which should be placed at God's service. "The more holy a woman is", wrote Léon Bloy, "the more woman she is." Which we can translate, "The more a religious has the qualities of her times, the better will she realize her vocation".

2

WOMAN TODAY

WE shall be talking about women in the fully developed countries: in Africa and Asia the situation is quite different. There, too, however, evolution is taking place and the trend seems likely to be irreversible. According to a recent U.N. report, only nine countries continue to refuse women the right to vote and to be elected.

CHRISTIANITY AND FEMINISM

Christianity started a new era for women. In revealing to the world the divine adoption of all the sons of God, Our Lord proclaimed at the same time the fundamental equality of the sexes and raised the dignity of woman for all time.

Reacting against the laws and customs of the Greco-Roman world, the Church in her turn proclaimed the equality of man and woman before God and in matters of morals. The extent to which this equality was disregarded in olden times when it came to judging questions of adultery or divorce, and the extent to which custom favoured man at the expense of woman, is well known. The Church, however, has always fought for the emancipation of women so that they should be accorded equal rights with men in the organization of their lives and the enjoyment of their freedom.

She also fought for the emancipation of unmarried women, giving rise to a flourishing of religious orders. By her respect for dedicated virginity she brought into relief the nobility of woman and her right to choose her own vocation.

In spite of the rough manners of that time, the chivalry of the Middle Ages also constituted a tribute to the dignity of woman and a mark of respect for her.

If we pass on from the Middle Ages to the Renaissance we note that as the de-christianization of society proceeds so the civil authority interferes more and more in matrimonial matters and introduces the idea of the incompetence of the married woman.

Later on, the civil law was to place more and more emphasis on this legal incompetence and to approve measures tending to the disruption of family life. Christians all too often tagged along behind their times, breathing without knowing it the air of paganism. Their conservatism led them to confuse authentic Christian tradition with human traditions which were no more than the product of their times and the anti-feminist prejudices thereof. At the end of a chapter on the feminist movement of the nineteenth century, Canon J. Leclercq wrote:

There is no doubt that the position of woman as established by the laws and customs of the nineteenth century was not entirely in conformity with her rights and her mission, but it is regrettable that in this feminist movement there were too few Catholics drawing their inspiration from the tradition of the Church. In general, Catholic intellectual circles adopted a defensive attitude consisting of defending as if it were Catholic tradition a civil code inspired by a man's world in the throes of becoming pagan. . . .

This explains why countries with a Catholic tradition . . . are those in which woman has advanced least, and why Catholic circles, which have great influence in the middle classes, often show so little understanding of the needs of women today.[1]

THE ADVANCEMENT OF WOMEN

However that maybe, the past is done with and we must now take into account what Lucien Romier called 'the Advancement of Woman'. He himself set this advancement at the end of the nineteenth century because he was particularly aware of the economic liberation of women. The date is in any case unimportant: what is important is that this advancement, which was achieved after a mighty struggle by women themselves, is an accepted fact and is continuing at great speed under our own eyes. To convince ourselves of this, let us indulge in a brief retrospect, as one does for example in the field of transport to show the amount of progress between Stephenson's *Rocket* and a Trans-Europe Express, or between Montgolfier's first balloon and an interplanetary rocket.

Then

A swift return to the past will permit us to assess progress in feminine emancipation. A comparison of the woman of today with her sister of 1700 or 1800 would be rewarding: certainly the picture would not be lacking in colour and picturesque detail. Nothing points up the evolution of customs more than a family album where one can compare our ancestors with their counterparts of today: hats and dresses are just one sign of the different modes of life in different epochs.

Then—a woman's life was spent almost exclusively in the home. Her existence was bounded by the three K's: *Kirche, Küche, Kinder* (Church, Cooking and Children). A respectable unmarried woman never went out unchaperoned. No woman could exercise any public office. Wel-

[1] *Leçons de droit naturel*, Vol. III, La Famille, Ch. VI, p. 395.

fare, just as much as administration, diplomacy, the universities, Parliament and a host of other fields, was the exclusive stamping ground of the male. Women did not vote. It is not long since academic authorities were against young women attending universities, and their admission to the faculty has only very recently been gained by a narrow victory in certain Catholic circles.

The idea of a lady ambassador, or senator, or air-pilot was just inconceivable. It went without saying that even for equal work there was no question of equal pay for women. A girl was not supposed to express an opinion in the choice of her husband, but to sit passively by while her parents arranged matters.

Woman lived imprisoned in a sort of immutable destiny, in the framework of an idealized archetype set up by men and remaining invariable. She was supposed to be docile, faithful, resigned, hardworking—but all within well-defined limits and sheltered from the draughts and winds of the outside world.

Now

A new type has been born—modern woman. She does not passively accept her fate, she takes charge of it. Freed from her former shackles, she evolves in an atmosphere that allows her to deploy her natural gifts. Greater life-expectation gives her, once her children are grown up, an extra life as it were. The culture available to her is wider and she has more leisure. All these new factors affect her position and her activities in the world and open them to almost unlimited extension.

A kind of feminine passivity, with life spent in retreat, has now given place to an ever more well-defined activity. Our industrial civilization has torn the woman from her hearth and opened the doors of factories and offices to her.

Two world wars mobilized millions of women behind the front where they successfully replaced men.

The classical picture in which initiative lies with the man, and woman's part is submission, is no longer current. Penelope at her weaving, Marguerite at her spinning-wheel, Juliet on her balcony, Sister Anne shut up in her tower living on expectation . . . nowadays they all smack of folklore.

There is nothing in this advancement as such that militates in Christian eyes against the subordination to her husband in the home demanded by St. Paul, but this subordination can no longer be validly carried over to the whole of life.

The advancement of woman is today a *fait accompli*, it has become a part of our customs and is accepted without question. Never before has woman's influence been so noticeable. Never has her psychological ascendancy been more pronounced. She plays her part sometimes even in the sphere of high policy. Thanks to the development of her personality and culture, she takes part in her own characteristic manner in the social, economic and literary life of the world. She no longer acts through man by her influence on him, but in her own right and under her own colours.

Women are to be found in Parliament, in the Cabinet, at the Ministry of Health, in scientific research centres, in the worlds of art and letters, and in the great social services. They have carved themselves an important place in the great technical field of communication media of press, cinema, television and radio. They write and speak and act, and their opinions and feelings are given their full weight in every sphere. We have seen new posts created, cut from the whole cloth to suit them in the widest variety of circumstances. This shows that they are wanted, accepted and effective. They are expected to humanize the social services and make them less bureaucratic and theory-ridden, and better able to adapt themselves to individual cases and the complexities of special circumstances.

THEIR AMBIVALENT ROLE

Women's contribution to our present civilization is considerable, for good as also, alas! for evil. Who can say what damage is done to mankind's moral conscience by women who publicize in the most shameless manner their nervous crises, their divorces, the amorality of their spectacular lives, their unbridled luxury and their disgust with life? People do not realize clearly enough the extent to which certain campaigns led by women with a great deal of noisy publicity—in favour of divorce or abortion or birth prevention—strike at the roots of family life and shake the most sacred values of Christian civilization.

Woman has the awe-ful choice of being Eve or Mary: she is rarely neutral. Either she ennobles and raises man up by her presence, by creating a climate of beauty and human nobility, or she drags him down with her in her own fall.

We can apply the words of Joseph de Maistre to her, "Man makes the laws, but families make the morals", since the heart of a family is the wife and mother. Woman holds the moral destiny of the world in her hand today more than at any other time in history. She makes it in her own image and makes an active contribution to the formation of public opinion in her articles, magazines and books. She takes part in the exchange of ideas and makes her influence felt in the drafting of laws as well as in social activity, whether it be in the field of education, medical care or leisure.

This unprecedented role of woman opens a new era in the history of mankind. A type of society which had imposed itself on mankind for thousands of years has disappeared never to return. A new type of society, already existing in the more civilized countries, will be adopted tomorrow by other continents where woman's evolution has been slower but is still inexorably continuing. The women of Africa and Asia will also look at the world with open eyes and play their part with their men in fashioning it. Such is the foreseeable course of history.

Lenin was able to write, "The experience of all movements of liberation proves that the success of a revolution depends upon the degree of participation by women".

This is a phrase not to be forgotten. Christianity is the greatest and most radical revolution for freedom in all history. The apostolate means nothing but the penetration of Christ through us to mankind and to society. The participation of women religious in the work of the apostolate can now be seen in high relief: their feminineness, with all that means today for their contribution, cannot be betrayed or stifled but must on the contrary be expanded and fortified by their vocation.

The place occupied by women in the world of today brings new dimensions to the apostolic work of contemporary religious. It remains now to consider this in greater detail.

3

THE RELIGIOUS IN THIS WORLD

How does the religious appear in the world of today in the present state of woman's evolution? That is the question we must now examine.

We are not concerned here with describing the religious through the eyes of an unbeliever. Without faith, how could he grasp the meaning of a life which exists only through and for Christ, in which everything is done in reference to Him? Without the faith which would solve the mystery for him, the religious life must appear an enigma, a waste of time, an inexplicable abdication of freedom, a closed door for which he has no key. It is but a short step from this to the conclusion that the religious life is a life diminished, less than human, based on repression and non-love; and it is a step easily taken, to judge from a certain stream of contemporary writing as tendentious as it is ill-informed.

What we are concerned with here is to see the religious through the eyes of a believer who regards her with sympathy and gratitude.

One thing that strikes anyone who is at all observant is the immense place occupied by women in religion in the vast field of human suffering. "Who suffers," said St. Paul, "and I do not suffer with him?" Women in religion *live* this phrase—at the bedside of the sick, with the handicapped and the bedridden, with old people, lepers, deaf-mutes, and prisoners. They live it, day-long and year-long, with a devotion and disregard of self which compel our admiration. They are standing witnesses to the Church's maternal love, concerned about all our miseries, mindful always of the parable of the Good Samaritan and Our Lord's words about anything done "for the least of these, my little ones". They are the vanguard of the missionary Church in jungle, desert or arctic ice: everywhere where there are missionaries they are to be found, perfect collaborators and assistants beyond price.

Another striking feature is the considerable place they occupy at all levels in the world of education. It is no rarity for non-practising Catholics to confide their children to them. People know that nuns have given up the possibility of a family of their own in order to be at the disposal of all families and be able to devote their whole care and attention to them. Masses of children and adolescents receive instruction and education in establishments run or inspired by them. People admire the nuns of these teaching orders for their watchful devotion and the unstinted trouble they take.

There are undoubtedly other ways in which religious are present to the world, for example in social assistance and parish work; but those we have mentioned are the most striking, and it is generally on them that the opinions and appreciation of the faithful are based.

It would, however, be neither objective nor complete to ignore a number of regrets that one finds expressed in innumerable articles, discussions and conversations. In general these remarks are not concerned with the essence of the religious vocation but with the overall customs and usages which make the religious seem in the eyes of the faithful to be living outside the world they are trying to save, to be lagging behind the general evolution of women.

Among these comments some deal with the psychological attitude of nuns, individually or as a community, while others are aimed at their social and apostolic attitude in the worlds of education and good works. Let us consider some of the more important ones.

PSYCHOLOGICAL ATTITUDE

Religious too often seem to be living in a closed world, turned in on themselves and having but tenuous contact with the world outside.

A community of nuns often enough gives the impression of being a fortress whose drawbridge is only furtively and fearfully lowered. A concept of separation from the world leads, people believe, to a kind of psychological isolationism, leading in turn to a failure of dialogue with those in immediate contact with them for lack of common interests and a common language. Even if a girl before her entry to the convent was thoroughly involved with the world on account of the circles in which she moved or even on account of her apostolic work, after a time she loses touch. She goes into an enclosure which, if it is not hermetically sealed, at least looks out on the world through openings more like arrow slits than bay windows.

Layfolk are familiar with the lost feeling experienced on coming back from a holiday away from home. Life has not stood still in their absence, changes have occurred which must be taken into account: in short, for a matter of days after their return they feel out of things and it takes some time to readapt. But religious do not resume contact: the break was made once and for all and the gulf can only grow wider.

Again, physical and psychological detachment from the world leads a religious to turn in on herself and her own community. Her world shrinks and if she is not careful will end up no more than a few square yards in size. Whence comes distorted vision, seeing everything from one angle, measuring things against a diminished scale. Whence, again, the contrived and artificial nature of certain customs in religious houses—a sort of 'house etiquette', a stilted, stereotyped and unnatural behaviour. It has been said of certain congregations of nuns that they are 'the last strongholds of the very studied manners of the middle-class woman of the nineteenth century'. People would like to see more spontaneity, less inhibition, more natural and straightforward reactions. It is not the respect and conventions they wish to see the last of, but the outmoded expression of them.

To put it briefly, the dusty old wax flowers should be replaced by living blooms drawing nourishment direct from the earth.

Most religious habits, too, seem to the layman to be ill adapted to current conditions, to have outlived their purpose, to be archaic and inconvenient. They raise at best an ironical smile when a nun is seen on her way to tend a sick person, flapping through the streets on her moped with her habit and veil streaming behind her to the imminent danger of herself and other traffic.

SOCIAL ATTITUDE

Apart from the foregoing comments which bear mainly on the individual or collective attitude of nuns, it has been observed that religious are in danger from officialism. This applies to nuns in hospital work and to those in schools. In the former of these two fields, State control of the health service and the growing demands of administration make it more and more difficult for a religious to carry out her proper religious mission as such.

The hospital has developed considerably since the days when foundresses of religious communities chose it as the field in which to bring succour to the crying distress of their times. In those days the care of the sick was first and foremost a matter of charity, and religious congregations were pioneers in the field of the corporal works of mercy. Today Ministries of Health have taken over much of the nursing and impose their own rules and regulations on what remains in private hands. The congregations have had to undergo certain changes by sheer force of circumstances. The necessity to have diplomas, the imposition of administrative controls and social legislation have all changed the view taken of the nursing nun, not by God but by our fellow-men.

She appears less and less as a religious bending over the beds of sick humanity, in close contact with her patients, having time to . . . have time. Hospitals, clinics, maternity homes all have large numbers of lay staff and the role of the religious is often no more than that of ward-sister or administrative superintendent; she is sometimes snowed under by purely administrative or supervisory tasks; she becomes more and more like a professional nurse overburdened with technical duties. She is in danger of becoming a mere official.

She bustles about and 'does' for her patients, but she cannot meet all their demands, for as the population increases so does the number of sick persons needing treatment. The poetry which enhalo'd the nursing nun of an earlier age when she cared for her patients in a directly personal relationship in some place like the Hôtel-Dieu of Beaune is sinking further and further into the past.

What is true in the field of hospital work, today so laicized, is equally true in the scholastic world. The Church has played a great part in the education and instruction of children, an historic part and one that was in part at least plugging a gap. The Church did not wait for the evolution of the modern State which would take charge of the formation of youth. She also undertook pioneer work in other fields not specifically within her competence. It is related that the Capuchins of Paris formed the city's first fire-brigade! But the past is past.

Today one cannot but notice the profound changes that have taken place in our educational establishments due to the growing demands of public authorities and the increase in the school-age population.

A network of laws governs private as well as public education. The teaching religious appears to lay eyes to be more and more just a teacher. The overburdened pro-gramme of work leaves less and less time for directly religious and apostolic contacts. Her pupils see her as a teacher who instructs them competently and devotedly in algebra or history and prepares them for their exams. The schoolmistress absorbs the nun in the eyes of the pupil and parent alike.

Here again, the danger of her sinking to the level of a mere functionary is by no means imaginary.

APOSTOLIC ATTITUDE

One cannot but be impressed by the sum of devotion represented by the religious life.

But devotion and the apostolate are not two realities which coincide. The question which everyone who has the care of souls should ask himself is this: great as their de-votion undoubtedly is, what apostolic return do the Church and the world get from the activities of religious today? Is there not perhaps some apostolic capital that could be put to better use? Are there not perhaps some fields of apostolic endeavour lying fallow, awaiting their cultiva-tion?

It seems to us that the answer to the questions must be, Yes. It is necessary, however, to be clear about what we mean by the apostolate properly speaking, and to define in advance in what ways devotion and the apostolate are different. This should make apparent what gaps there are to be filled.

Let it be clearly understood once and for all that in speaking of gaps we are by no means levelling any criticism at the religious of our times. It is not their fault if they have not been trained for nor asked to undertake certain missions that are necessary today.

We shall try later on to unravel the causes of this situation.

MEANING OF THE APOSTOLATE

Having said that, we must also take into account the fact that though the apostolate is as old as the Church itself, it is a duty that has become rather shadowy in Christian consciousness, partly because people have thought they were living in a Christian society, that is in a society where the non-believer was the exception. The revival of the apostolic conscience, and in particular the awareness of the need to organize it, are recent happenings. The last few Popes have issued urgent appeals reminding the faithful of their apostolic duty inherent in their baptism and *a fortiori* in the taking of religious vows. But the idea has not yet really penetrated our everyday views or the constitutions of religious communities.

It is of the greatest importance to insist on the existence of this duty and to clarify its meaning, that is to say to make it quite clear in what the idea of apostolate consists.

'Apostolate' is a word which is used in different senses which have evolved in the course of time and which do not always embrace the same content of meaning. We shall use it here in its 'missionary' sense to denote the activities of the Christian sent by virtue of his baptismal vows to bring Christ to the world, either by revealing Him to those who do not know Him or by increasing His empire in those who know Him already and training them to preach Him in their turn. In both cases it is a matter of a supernatural communication of life to bring Christ into souls and into the world and to foster His increase. It means working for the increase in extent and intensity of the Kingdom of God here below. The apostolate is the work of evangelization, either of spreading the gospel or of making it penetrate deeper into all human and social activity. This apostolate is something we have to do ourselves and train others to do after us: doing and making others do are integral parts of the same whole. In the following pages special emphasis is laid on the duty incumbent upon women in religion to draw into the apostolate those lay persons with whom they are in closest touch.

This does not mean that the life of religious devoted to education, care of the sick, or social work is not already apostolic: it is, by reason of their consecration to this work. As we said in the Foreword, it is not a question of the advancement of religious from a non-apostolic to an apostolic sphere, but it does mean that certain apostolic factors have to be introduced into the very heart of the religious life.

The tendency nowadays is to minimize the religious side, properly speaking, of the apostolate which tends to be overlooked in a welter of things which prepare for the apostolate, ensure favourable conditions for it, support and extend apostolic action, but are by no means to be confused with it. We are not concerned here with clearing the religious apostolate—in the sense of the Gospels and the Acts of the Apostles—from all that which, seen from that angle, is but pre- or para-apostolic. Suffice it to say that the apostolate is the extension of Christ's mission in and through the Church, a mission which consists of giving God to the world, of acting in such a manner that men come to know God, to love Him and serve Him, to take their nourishment from Him, and to live the whole of the Gospel in every aspect of their whole lives.

The apostolate, then, in its character of evangelization is not to be confused with works of dedication, however necessary these may be.

There is a tendency also to use the term 'apostolate' for activities which are apostolic only in intention. This does not mean that this intention is without value or significance; merely that by stretching the application of words one robs them of any real meaning, so that the original sense is lost. Let us be quite clear that any action whose final aim is to glorify God has a supernatural and redemptive value, but this is not to say that it is intrinsically apostolic. The word 'apostolate' is as much abused as the word 'prayer'. One hears it said sometimes that work is prayer. It is and it isn't. One can work in a spirit of prayer, and a very good thing too, but one must not confuse two separate values. We should add that in order to make a 'prayer' of one's work one must have acquired elsewhere the meaning of prayer. In the same way, if one wants to acquire the apostolic mentality that can inform even a course in geography it is extremely desirable to have acquired, by practice of the apostolate properly speaking, the proper orientation and perspective of life.

Given these definitions, it is easy to show in what way works of dedication are distinct from the apostolate. One can dedicate a whole life to the care of the sick, but one does not begin to be apostolic until one confides in them the secret of one's devotion, until one leads them to know Christ and does one's best to make Him loved. Devotion opens another's soul to sympathy and favourable prejudice and disposes him to lend an attentive ear to what you say, but so far it is not passing on the message or communicating life.

A French bishop, Mgr. Huyghe, speaking about a community of nuns nursing in people's own homes, which had been established for thirty years in one particular town, gave the following account of what one of the nuns told him:

"We are now in touch with the third generation. When we came we knew only the old and the sick; we nursed the next generation; today it is the grandchildren of our first contacts who come to us for help. We have become completely naturalized in this working-class quarter and we are welcome in every home. We have, of course, been able to help bring the priest to the bedside of the dying in many cases, but we have never succeeded in converting one single adult in good health. To anyone with eyes to see, it is clear that our quarter continues to become more and more de-christianized."

To which the bishop added this remark which is worth remembering: "One can dedicate oneself and still not reveal the person of Christ."[1]

This quotation turns the spotlight on a state of affairs that we cannot accept as normal and against which we must react.

We shall have occasion later to examine the real apostolic return from religious in the spheres with which they are already familiar. For the moment, it seems important to draw attention in general terms to a field of vast extent which remains fallow.

THE ADULT WORLD

An observer analysing the part played by religious today cannot help being struck by their absence from the main spheres of influence at adult level, spheres where they have a right to be and where their talents are called for and their presence is needed.

[1] Quoted in *Equilibre et Apostolat*, p. 233.

As a general rule the life of a religious is dedicated to children, to the sick or to the elderly. No doubt it is possible to reach the family to some extent through the children or the sick, but such an influence is necessarily indirect and often sporadic. The grown-ups as such are outside their influence. Nevertheless it is grown-ups who run the world, create the climate of opinion and the atmosphere we all breathe. It has too often been said that to form the young is to assure the future. This is true, but only to the extent that the formation continues until the young adult goes out into life and founds a home; it is true, but only to the extent that the influence on youth is complemented by action on the adults who in their turn will form, or deform, the coming generation. The self-perpetuation of schools is a myth that has sometimes cost us dear. One recalls the remarks of a notorious communist about our schools: "We leave you the children; we take care of the grown-ups." Or again: "Teach them to read and write; we'll teach them to think." This surely is invitation enough to carry our work through to its logical conclusion. It is seldom that one sees religious playing any part at adult level, a level at which other women, however, make their influence felt, baneful as that influence so often is.

One does not see them, either, playing any part among adult lay-women whom the Church calls to the apostolate but who often lack anyone to stimulate and sustain them. One meets religious at various Catholic Action congresses which take place in their houses during the school holidays, but it is only in the kitchen or the refectory or perhaps at the closing session that they meet the public and the guests. One rarely sees them play any real part among young Catholic Actionists who, though they have just left their schools, seem to expect nothing more from them.

In most of our university cities nuns run halls of

residence for the female undergraduates. With rare exceptions their work is restricted to being hotel-keepers, housekeepers or cooks. It is the exception that they manage to put some real life into these homes where tomorrow's adults are being shaped, or to produce any effective spiritual guidance for them at this decisive stage of their career. All the girls expect from them is board and lodging —and the greatest understanding about their freedom to come and go at will.

It may be argued that the youth in question hides itself away and is jealous of its freedom. True. But it is also true that our instructresses have not had the training to enable them to maintain contact with the young adult woman and remain close to her at a time when she has to make choices and decisions that will affect her whole life.

In many towns and pilgrimage centres religious devote their lives to running homes or boarding-houses and are exclusively concerned with the domestic affairs of running an hotel. How can one fail to deplore the fact that they have no apostolic outlet for their abundant spiritual energies? They have consecrated themselves to God and to souls for this work no doubt, but principally for something quite different. Such employment of nuns devalues the religious vocation in the eyes of the faithful. Priests, too, are worried to see so many dedicated souls harnessed exclusively to material tasks when there is such a crying need for all persons of goodwill in directly apostolic work. They, too, cannot but deplore the fact that the apostolic return of such religious is not even equal to that of many lay people in the world.

It does sometimes happen that an individual religious, however dedicated, is unable to devote herself to apostolic work. Nevertheless, the community as such ought to undertake it. The religious who, for some special reason, cannot play a direct part in the work can make her contribution to the apostolic work of the community by carrying out her secular tasks for the love of God and souls. She will thus enter into the great apostolic current so much desired by the Church. But her case should remain the exception.

Let there be no misunderstanding. We are not criticizing the obscure jobs which lack glamour, but the fact that apostolic activities of far greater importance are on this account deprived of their help. Obviously a truly apostolic soul does not prefer one job to another on account of its external glamour but for what it does, or makes possible to do, for souls. The whole matter is summed up in the question: where and how are the cause of God and the visible extension of His Kingdom best served?

4

UNEASE AND DEVALUATION

HESITATION ON THE THRESHOLD

THE picture we have just painted comprises highlights as well as shadows. The positive contribution which religious can make is so great that it ought by itself to supply the deficiencies. Yet it remains true that the deficiencies, and the element of truth in current criticism, are hampering recruitment. Everywhere one hears complaints that recruitment does not keep pace with needs, nor indeed with the growth of population; that religious houses are closing down one after another; that the average age of communities is increasing. This is as much as to say that the vitality of the Church is affected in her live forces, in this élite of womanhood that is at once her glory and her means of exercising in a special way her role of spiritual motherhood.

Vocations are decreasing everywhere. It is significant that the least affected are the missionary and purely contemplative congregations. The ideals of heroic missionary work and of a life devoted to God in silence still appeal to youth. On the other hand, vocations in hospital and teaching congregations are falling off to some extent everywhere.

A Belgian bishop has drawn attention to the fact that in his diocese out of 522 religious houses of women 78 had to close in the last thirty years for lack of new vocations.

In one French diocese the numbers of religious in education and nursing dropped 30 per cent. in ten years.

On top of this comes the factor of ageing communities.

The age pattern of many of them makes it clear that the rate of recruitment is falling off. The Bishop of Tournai has revealed the following figures in a pastoral letter: nearly 30 per cent. of nuns are over 65 while only 10 per cent. are under 30. This factor of ageing communities will have to be taken into account when it comes to considering modernization and the adaptations necessary.

The falling off in religious vocations is serious in itself, but it would be less so if it were purely numerical. What is much more serious for the future is a sort of spiritual devaluation of the vocation which is noticeable among good Christian families, the lay apostolate and even among the clergy.

Among the many causes of this devaluation must certainly be ranged the de-christianization of society and its customs, the increasingly materialistic atmosphere in the world and in ourselves, the internal disorder and egoism lying under the surface of so many families, and the decrease of the spirit of faith. . . . But, rather than list those reasons which lie outside our own responsibilities, it would seem more useful and more efficacious to seek the causes which can be placed at the door of the religious congregations themselves and which have been laid bare by a series of enquiries.

The complaints—even if they are not always and everywhere applicable and even if they sometimes underestimate the positive contribution of the religious congregations— are too numerous and too similar not to warrant the most careful study.

Vocations, we said, are on the downgrade in the teaching and hospital congregations. It is important to understand the psychology of a young woman hesitating on the threshold of these vocations.

If she is thinking of a teaching congregation, it is obvious that what attracts her is not the idea of teaching geography or algebra or the domestic arts; what she wants is to become an apostle of Christ, to bring God to the world in and through such and such a congregation. The attractive power of a congregation lies in its apostolic value. Even if the girl is not herself fully aware of it, what decides her is the picture of a religious vowed to God and able to win souls for Him.

It is the same if she is considering a hospital congregation. What attracts her is not the profession of nursing as such—the time is long past when one had to become a nun in order to tend the sick. She weighs the pros and cons of a lay nursing vocation and if she then decides to enter religion it is in order to be able to give the sick, and through them the world, not only nursing care but also an overflow of Christian life and of happiness. She wants to be quite clear about the difference between being a lay nurse and being a nurse in religion, and it is necessary for the plus-value of the religious version to be quite clear before she will choose it. All studies of the problem of vocations should be informed by this fundamental truth.

UNEASE WITHIN

Nuns themselves experience in a different but none the less real degree the difficulty of reconciling scholastic or medical exigencies with the spiritual and apostolic aspirations which formed their vocation. All Christians live a paradoxical life, since Our Lord demands of us that we should be in this world but not of it. Religious live this paradox in community, bound by a very definite Rule. It is not surprising that from time to time a problem arises on how to maintain the delicate balance between the demands of their spiritual, apostolic and professional life.

It often happens that a community has difficulty in finding a suitable compromise between the Rule which separates it from the world and its apostolic work which demands certain contacts with the world.

Instead of ignoring the problem, which is by the grace of God perfectly soluble and has been satisfactorily solved by some communities, one must see how to reconcile what must be reconciled.

The Church values both the interior life and the apostolic life, separation from the world and the action of leaven in the world. One has only to take literally the repeated appeals she addresses to her favourite children, her religious, to see all these difficulties give place to harmony.

Everybody realizes the awkwardnesses involved in adapting the religious life to the world of today. The first to realize it are the religious themselves who have often experienced a conflict within their own communities.

Young religious who have been in touch with the contemporary world are aware of the values appropriate to our time. They feel that certain customs of the religious life no longer fit in. A more direct and less inhibited manner, a wider human culture, and above all experience of apostolic movements before their entry into the novitiate, have opened new horizons to them and given them a sense of responsibility. They come to the convent not to give less of themselves but to give more, and they are very sensitive to anything which puts the brake on or lessens the apostolic ardour with which they enter. They want a religious life which, albeit different in expression, is yet of a piece with their immediate past. Young religious also expect a spirituality richer in biblical life and communal liturgy, less dependent on 'spiritual exercises' and non-liturgical prayers. It is important to take this into account. This conflict between the generations can be softened by the understanding of Superiors in many cases, but it exists, even if latent, throughout.

D

To this kind of un-ease among the young must often be added that of their elder sisters who are conscious of the need for evolution. For the better ones among them this may constitute a case of conscience : on the one hand they realize the need for change, on the other a certain conception of obedience urges them to stick to the established order of things, to abide by the *status quo* which does not question accepted uses and customs. Seen in this light, any innovation seems doomed in advance. Authority easily becomes a sort of gentle and maternal authoritarianism, and obedience becomes no more than a passivity which solves problems by ignoring them.

Another difficulty is often added by differences in apostolic perspective between a mother-house in Europe and daughter-houses overseas. The latter have often been driven to make the necessary adaptations resulting in comparisons to the disadvantage of the European houses.

These factors combine to bring about a most unfortunate result—the devaluation of the religious vocation in the eyes of the world. Let us pause a moment before this saddening picture.

DEVALUATION ON THE OUTSIDE

It is an undeniable fact that the wind of defeatism is blowing about the older, traditional congregations. It is widely held that they are all condemned sooner or later to extinction, that history has already passed them by, as has happened to so many previously flourishing congregations. A whole mass of popular literature underlines the pessimistic view of their situation. M. Baldwin, K. Hulme, A. Hure, F. Werfel, G. Walschap and even, after their fashion, Bernanos and Montherlant, help to sustain the prejudice against their survival. It is no exception to hear it maintained, even in our own Christian circles, that this is the hour of the secular institutes or that the world needs entirely new forms of the religious life. The tendency is to discourage those who want to renew the religious life in its traditional form : they are reminded about putting new wine into old bottles, told that tradition has become top heavy and immobilizing, that they must resign themselves to seeing congregations which are out of step with contemporary history plodding painfully along on the fringe of the world's development.

The religious of today appears to the faithful to be out of touch with the world as it is, an anachronism.

She seems also to be behindhand in relation to other women, who have achieved emancipation while she remains 'in the schoolroom'. To revalue the religious life of today means, therefore, to bring the religious life into harmony with the evolutionary state of the world and womankind, to retain from the past everything of lasting value that can be adapted to circumstances, and to accept the positive contribution of feminism in order to improve the apostolic yield.

The process of revaluation means also synchronizing in some measure the religious life with the continuing evolution of the world. The position of a religious in the world today ought to be defined in terms relating to this world. To the question "What is a religious?" one ought to be able to answer, "A contemporary woman—not one of the eighteenth or nineteenth century—who has dedicated her life to God for the salvation of the world through the congregation to which she belongs."

Every phrase in that answer has its importance.

She must be seen to be a contemporary woman.

It is therefore essential that she should have her proper place in the development of ideas and customs that distinguish the woman of today from her sisters of a century or even half a century ago. Anything in the life of the religious which does not fit in with the present state of feminist evolution is a hindrance to her apostolic activity.

She should appear as a woman vowed to God in the Church and in the world of today.

Her vows bind her to God—this is valid at all times and constitutes the unchangeable element of her vocation—but they also bind her to the salvation of the world in which we live. The addition of this rider immediately makes it clear that she must introduce her apostolic work into the world *as it is* and that there is a continuing task of adaptation to the exigencies of the moment; and even more, that she must fit herself into the vital currents which actuate the Church in which she lives.

This dedication to God for the salvation of the world must be implemented by each religious through her own congregation. She is thus placed on her proper path, a path which is defined by her Rule, which must be faithfully adhered to. When we say '*through* her congregation' we are making the point that the congregation is a means to a more important end, an end which draws it, as the sea draws a river, into the common stream of world evangelization.

It only remains for us to examine in detail what this fundamental definition requires if we are to retrieve the situation and hasten a renewal which the world and the Church so urgently need.

What we have said, it must be emphasized, is the very opposite of defeatism. It has been inspired by complete and absolute faith in the value of religious congregations of the classic types. Without undervaluing the providential part played by secular institutes or that played in the world by the different forms of religious dedication which the Holy Ghost never ceases to raise up, we believe that it would be a grave fault not to do all in one's power to reinstate the classic religious vocation at its true value.

Before seeking remedies for the present situation, it is necessary to analyse it more in detail. What are the causes of the disharmony we have noted? What is their origin? It is to the examination of these various factors that we shall now proceed.

5

CAUSES OF THIS SITUATION

It is essential, if we are to have a renewal, to resolve the uncertainties underlying this predicament and to establish the causes which led to it. It is due in great part to a lack of harmony between the practice of religious life and the demands of the apostolate. How can this dualism have arisen, since the religious life is no more than a fuller version of the life we are called to by baptism and the duty of the apostolate is inherent in baptism? To appreciate this situation, at once paradoxical and abnormal, we must consider a great many different factors working together to produce the uneasy situation. Once the main causes are known, the remedies will be obvious: diagnosis is already one step on the road to a cure, a cure which is within our powers if we so wish.

For reasons of clarity we shall divide the various factors behind the present situation into historical, spiritual, canonical, psychological and sociological. Let us examine each of them briefly in turn.

A. THE HISTORICAL FACTOR
'Religious' synonymous with 'enclosed'

Anyone who knows the history of the various orders and congregations is aware that to start with, and for hundreds of years, the concept of 'religious' was applicable only to contemplative orders.

A religious was by origin and by definition an enclosed

person taking solemn vows. The idea of unenclosed nuns seemed a profanation. Slowly the active congregations had to fight their way, not without struggles and misfortunes, to gain the right to be included in the term 'religious'. The history of the cloister down the ages has much instruction to offer on this point.

Evolutionary Stages

Whereas deaconesses had played an important part in the primitive Church, including giving Holy Communion to sick women, nothing approaching this was permitted later on. Once the deaconesses had disappeared during the fifth century, the Church knew only the enclosed monastic life for the next thousand years. The most intransigent manifesto in favour of enclosure was the celebrated decretal *Periculoso* of Pope Boniface VIII in 1298:

> We command by this present constitution, whose validity is eternal and can never be questioned, that all nuns, collectively and individually, present and to come, of whatever order or religion, in whatever part of the world they may be, shall henceforth remain in their monasteries in perpetual enclosure.

We had to await, alas! till the beginning of the sixteenth century to see a relaxation of the rigidity of the *Clausura*. It came at the end of a long conflict of trends, between jurists and pastoral clergy, and led eventually to a certain apostolic 'promotion' for female religious.

The revolution took place in the second quarter of the sixteenth century. The impetus came from those male religious who abandoned the classic type of cloistered life and made their immediate aim the apostolate. It was typified by the Barnabites and, above all, the Jesuits: and it was to have its repercussions among the women.

Until then one had hardly conceived the idea of religious

not subject to enclosure and taking solemn vows. The 'apostolic' innovation was to be resisted and to experience various difficulties which are instructive for us today. At the front of this battle was Angela Merici. In 1544 the Ursulines founded by her received approval and were not subject to enclosure. Her first twelve sisters were dispersed and accommodated each in her own parish. They devoted themselves in various quarters to a multiplicity of works, but especially to the education of young girls and particularly those of the common people whom no one else worried about. Unfortunately their freedom of movement did not last. They were obliged to accept grilles, and shortly afterwards, in 1566, the Holy See ordered the suppression of all female congregations not in enclosure and subject to solemn vows. The Ursulines, who up to then had dressed like anyone else, were required to adopt a religious habit. It was the triumph of legalism.

In the footsteps of St. Angela Merici, St. Peter Fourier and Bd. Alix Le Clercq tried in their turn to avoid enclosure. The founder of the Canonesses of St. Augustine was quite clear about what he wanted: he wanted to establish free schools for day pupils and boarders and he wished the exigencies of the education of children to take first place over the rules of enclosure. "I have always thought", he wrote, "that it was necessary to say that they were first and foremost schoolmistresses . . . for fear they should be thought to be religious first and foremost."[1]

Canon lawyers made it tough for him. To add to his misfortune, Urban VIII rejected the idea of teaching nuns on account of a passage in St. Paul: *docere autem mulieri non permitto*. In the end the work was approved, not without trouble and worry, but within the classic framework.

At the dawn of the seventeenth century St. Francis de Sales, with St. Jane Frances de Chantal, tried to promote the

[1] E. Renard, *La Mère Alix Le Clercq*, p. 292. Paris, 1935.

apostolic activity of women in the world. He was thinking in terms of an unenclosed congregation, not taking solemn vows, at the service of the poor and the sick and visiting them in their homes. The Order was formed under the title of Our Lady of the Visitation. But the Archbishop of Lyons came down on the side of the legalists, and despite encouragement from St. Robert Bellarmine, who was consulted and who advised him to stick to his ideals, St. Francis bowed sadly before such deeply rooted tradition.

The battle lost by St. Francis de Sales was to be won by the exercise of patience and 'holy cunning' by St. Vincent de Paul. In order that his Daughters of Charity should be able to leave their convents to serve the poor, he used all his gifts of diplomacy to get round the rigours of canon law. To avoid enclosure, obligatory for all female religious, the institute was called a company and the novitiate a seminary; the superior was designated Sister Servant, and their residence was called not convent or monastery but just 'house'. He did not talk about religious but about 'daughters of the parish'. Aspirants were not given the veil, but kept the *toquois* for headdress and wore the grey serge dress of the common people.

He left instructions, famous today but very daring at the time. Read what he writes, and repeats, to his daughters:

Should the local bishop ask you if you are in religion, you will say that by the grace of God you are not, not because you have not a high opinion of religious but because if you were you would have to be enclosed and that would mean goodbye to the service of the poor. . . . Should some muddle-headed person appear among you and say, "We ought to be religious. It would be much nicer," then, my dear sisters, the Company is ripe for Extreme Unction, for who says 'religious' says 'enclosed' but the Daughters of Charity must go everywhere.

In the Rules he wrote:

> They shall consider that they are not in religion, since that state is unsuited to the tasks of their vocation. The Sisters have no convent but the houses of the sick and the house where the Superior lives, no cell but a hired room, no chapel but the parish church, no cloister but the streets of the town; for enclosure they have obedience . . . for grille, the fear of God; for veil, holy modesty.[1]

Such, in brief, is the history of a battle fought for the freedom of the Holy Ghost. St. Vincent de Paul made an important break-through but he did not cover the whole ground. In the centuries separating us from him evolution has been going forward, as it still is, and the Council may well take it a decisive step further.

This historic sketch helps us to realize how it happened that laws suitable for the contemplative life came to be imposed for so long on those who were trying to carve out a path for active congregations.

The Spirit of the Times

To this must be added that religious constitutions did not escape the spirit of their times and were marked, as other Church institutions were, by the ideologies and deficiencies of their epoch.

Those dating from the seventeenth century were greatly influenced by the viewpoint of the Council of Trent, which was much concerned with laxity and internal abuses. Churchmen were thinking in defensive terms, not of attacking; they were concentrating on what was going on within the household more than on the world outside whose dechristianization had not yet come out into the light.

The eighteenth century, overshadowed by the French

---

[1] St. Vincent de Paul, *Correspondances, Entretiens, Documents*, Vol. IX, p. 533; Vol. X, pp. 658, 667, 662. Librairie Lecoffre, Paris, 1923.

Revolution which was to bring it to a close, was not a century of religious expansion.

On the contrary, the very foundations of community life were in danger in a struggle for existence. It was only by counter-action that the Revolution, by demonstrating the inconvenience of the existing laws about vows, contributed to the development.

As regards the constitutions drawn up in the nineteenth century, one must not minimize the influence in our countries of certain remains of Jansenism which placed great emphasis on the withdrawal from and the hatred of the world, and on the intrinsically corrupt nature of man as the result of original sin, on the fear of God divorced from love, on the contagiousness of sin, on the repression of natural tendencies. One has only to glance through some of the writings of the time to see the traces of this tendency, which only gradually died away.

If one also takes into account the fact that for centuries woman played no part except in the home and behind the scenes, one can understand why it was impossible to stress the idea of a religious woman 'in the world', since all women were in the background. Feminine emancipation put an end to this, but the results of this emancipation of women have not yet been fully felt in the organization of religious life. To the great loss of the apostolate of religious, religious life has lagged behind in the vast field of action opened to the direct influence of women by their emancipation.

As far as more recent constitutions are concerned, one must remember that the idea of an organized lay apostolate only crystallized into Catholic Action under Pius XI. The idea that every baptized person should be an apostle is slow to take hold in the Church. With this in mind, one cannot be surprised that the concept of vows of religion being an extension and fulfilment of baptism has not been sufficiently publicized, and one understands why the role of religious as the moving spirits of the lay apostolate among women has not even been touched upon in these constitutions.

In the nature of things one breathes the air of one's times and lives within their limits. Once more, it is not the fault of the religious, who are quite properly intent upon obedience to their Rule as it is and are not called upon as individuals to make adaptations of it.

B. THE SPIRITUAL FACTOR

Monastic Spirituality

The contemplative origin of religious orders explains in part why apostolic spirituality has not had the full development it merits. A purely contemplative spirituality has had gradually imposed upon it a spirituality more nearly directed at action, but the balance between the life of prayer and the life of the apostolate has never been fully attained at the spiritual level itself. One slips very easily from one to the other—or rather, the 'contemplative' aspect retains a primacy which on some points fits in badly with the very real exigencies of the active vocation.

First of all one must distinguish between contemplation as such and the contemplative life. Contemplation, that is prayer in its purest form, is a part of every religious life and should have a very important place in it, but the contemplative life means something quite different—a life arranged on the basis of complete aloofness from the world. The vocation to which it is the answer is of the very highest value, but it is not the vocation of congregations with directly apostolic aims.

The contemplative life has been described, quite properly, as seeking God principally in Himself and for Himself. It corresponds to the duty of direct adoration and is centred on the liturgical life—*Opus Dei*—and on the virtue of religion. The apostolic life, on the other hand, is oriented towards God in Himself and the service of God in serving one's neighbour and bringing him to love God. In a famous phrase, the apostle abandons God for God's sake. The phrase is, incidentally, inexact, for the apostle does not abandon God but remains in communion with Him while serving Him through his neighbour and adoring Him through His creatures in whom He abides.

Again, for those vowed to the contemplative life, as for those confined to a bed of sickness, the apostolate has a different meaning and is exercised in a different way. The contemplative life is eminently apostolic in the intention which inspires it, regardless of there being no visible active collaboration in the extension of the Kingdom of God.

The silence and prayer of contemplatives bring down graces. But these graces need active helpers if they are to be fruitful. God needs man; He needs our active collaboration just as He needs wheat for the Holy Eucharist and water for baptism. Grace normally moves in human channels ever since God became man and intended the Incarnation in a certain sense to be continuous.

The religious who is vowed to the apostolic life must extend her prayer by action on every possible occasion. For her it is not enough to take refuge in prayer—to 'pray for poor sinners'—in order to be dispensed from the action which might perhaps help them to get free of sin. The primacy of prayer, any more than God's omnipotence which can work all the miracles of grace unaided, cannot serve as an excuse for inaction on our part. The religious vowed to the apostolic life has her own way of interpreting the imperatives of prayers. "The temptation of active religious", a Superior-General wrote to us, "is to want to become enclosed nuns and to organize their religious life as they do."[1]

---

[1] Cf. *L'Eglise en état de mission*, Ch. III, L'Excuse mystique.

What is proper and useful in one sort of life, then, is not necessarily suitable for another : words written for a Carthusian monk will not apply without some modification to a Sister of Charity. Let us take an example. Everyone knows the unparalleled position held by the *Imitation* in the spiritual life. This book, which has with good cause nourished many generations of nuns, contains a whole treasury of wisdom—but one must know how to read it; it was written by a contemplative for contemplatives. This accounts for the insistence with which it harps on the importance of solitude and aloofness from the world. "Every time I go back among men, I feel less a man myself." "It is praiseworthy for a religious to go out but rarely, to avoid seeing or being seen by men." But none of this is true if one goes out, not for personal satisfaction but to look for the lost sheep or to serve God through mankind. The fact that the *Imitation* has nothing to say about the apostolate is fair indication that its orientation is quite different. The book, which was written as a reaction to the danger of worldliness threatening a religious who had chosen the contemplative life, is to be read quite differently by one whose life is dedicated to the visible salvation of the world and who, for that reason alone, must in some way be in touch with people. Underlying all appears the conclusion that it is more worth while to withdraw from one's fellow-men and devote one's whole time to prayer. This is in conformity with the overall idea of the vocation of a monk vowed to silence, but it is not true without modification for others.

C. THE CANONICAL FACTOR

To the historical and spiritual factors we must now add the canonical factor. Canon lawyers, like anyone else, are men of their age. In the course of the years they have codified the religious life on the basis of the cloistered type of nun, and in the spirit of an age which treated woman as a minor to be protected from herself. Canon law still bears the marks of this masculinist mentality which has not yet entirely died out. It is well known that what one can only call the anti-feminist tradition has had a long innings.

It goes all the way back to Tertullian whose influence in the matter was powerful. Mindful of paradise lost through Eve, he wanted to keep women in subjection and clothe them in garments of mourning and penitence. Addressing women of all times, Tertullian cried out, "Do you not know that you are Eve? You are the devil's doorway. It was you who profaned the Tree of Life, you who dragged down with you him whom the devil dared not attack direct. It was you who thus disfigured the image of God which is man."

Ancient canon law reflects this unfavourable prejudice.

In the decretal of Gratian there are the following unequivocal passages:

> Woman was not made in God's image . . . so one can understand the desire of the law that women be subject to men and wives almost the servants of their husbands.

> It is clear that woman is under man's dominion and has no authority, nor can she teach, give evidence, make a contract nor be judge.

Some theologians, like some preachers, have fallen into step with this idea and tried to make woman into a sort of perpetual minor. Even St. Thomas followed his master, Aristotle, somewhat too unquestioningly in this matter.

"In her particular nature woman is something defective and accidental. . . . If a girl child is born, it is due to weakness of the generative principle, or imperfection in the pre-existing matter, or to a change produced by external causes, for example by the humid winds from the South, as Aristotle says."

That eminent Roman canonist, Father van Biervliet, consultant to the Congregation for the Affairs of Religious, having emphasized the anti-feminine attitude of the ancient canonists, came to the following conclusion.

> It is not surprising that canon law long reflected the common conviction of woman's weakness and her incapacity in many matters. From this came the dominant concern to protect her and supply her deficiencies by men. A typical case is that concerning the enclosure of nuns. For centuries the adage *aut maritus aut murus* (a husband or a wall) was the accepted principle. To defend woman against the attacks of a brutal and not very well organized world there was no other solution conceivable: a wife was protected by her husband, if need be by cold steel; but an unmarried woman concerned for her virtue had to be shut away. True, a woman is still exposed to danger, but in our modern society immorality is less open and it is easier to avoid it behind the protection offered by the regular constabulary and an ever-alert police.[1]

It would be easy to find many traces left by an unconscious anti-feminist attitude in the summary judgments given by canon lawyers and spiritual writers on the subject of feminine psychology.

On the subject of this hard-dying prejudice, a Superior-General once wrote to us:

> One feels that some measures were taken in previous centuries by men who mistrusted the uneducated woman. It is worth noting also that the rules were often the same for male religious [she added] but men are

[1] *Regina Mundi*, No. 5, 1956.

cleverer than we are at getting at the spirit of a text and do not allow themselves to become prisoners of the letter.

The naturally contemplative tendency of spirituality, reinforced by a series of canonical prescriptions, tends to inhibit rather than to favour apostolic zeal. It is only in recent times, since the repeated appeals of H.H. Pope Pius XII urging on all religious, including contemplatives, certain forms of apostolate, that we have seen more supple adaptations appear. The process of broadening the canon law is under way, but it is far from complete.

D. THE PSYCHOLOGICAL FACTOR

Astonishing as it may seem, among the causes of the situation we have described must be emphasized the deviations which threaten the practical application of the vows. Their vows are of sublime greatness, as the Church has often had occasion to repeat through the ages in reply to those who, every now and then, attack them as implying an alienation incompatible with human dignity and freedom. But their very greatness demands that they be interpreted in action down to the last detail with the purity of intention and delicacy of touch belonging to all things divine. Sensitive handling is necessary for all that concerns the depths of the soul. A fall on level ground seldom causes broken bones, but if one stumbles at a great height the risk is much greater: the height itself demands greater precautions and clear vision. And this is greatly to the honour of the religious vocation.

There is danger of distortion as soon as one adopts a too negative interpretation of the vows; that is to say, when the emphasis is on detachment and renunciation rather than on the positive aspect of cleaving to God. A one-sided emphasis tends to confuse the means with the end, and to present the vows as an end in themselves when they are in reality only the means to a full expression of the love of God.

E

Let us confine ourselves in this connection to one aspect of the practice of the vow of obedience.

It has been rightly said that the vow of obedience is the very essence of the religious life. By itself it can attain even the object of the other vows and it embraces the whole of the religious life, unifying it in a basic contract with God through the authority which represents Him. Every religious understands its meaning and its value. The more fervent she is, the more she wishes to anticipate the orders and even the wishes of her superiors. She does not therefore allow herself to question anything; in any case, feminine psychology gives her a tendency to docility and passivity. It is natural for her to obey: she easily accepts general directives, but will want to discuss the details, whereas man more easily calls the principle in question and will not be bothered with details.

The religious quite properly regards the orders of her Superior as an expression of God's will, and she does her best to renounce her own will and forget her personal preferences. If she is not asked in the name of obedience to undertake apostolic works or to open her heart to the broader apostolate which we shall describe, she has a duty to refrain from them even though she may be conscious of their necessity and urgency.

The canon lawyers have framed her life within clearly circumscribed limits, which she respects even if she feels she is being subjected to a kind of canonical tutelage. She accepts the fact that she can be made use of without her own wishes being consulted; she does not complain because, quite apart from the vow of obedience, she is encouraged to do so by a natural subordination which makes her content to lean on masculine support. Nothing in her make-up suggests that she demand the freedom which men in religion enjoy. The exigencies of the priesthood doubtless account for some of this freedom, but not for all of it, and the differences between nuns and priests remain very marked. As a nun she will not allow herself to claim even a few of the rights that woman has managed to obtain bit by bit from man in the world.

All this shows that if the Rules themselves are not opened to the new apostolic perspectives which we are going to describe, the vow of obedience will prohibit in advance any evolution generated from within.

E. THE SOCIOLOGICAL FACTOR

Finally, let us look at the sociological factor which has made profound changes in the field of action of our congregations, that is in the scholastic and hospital fields. The increasing socialization of life and the growing State control in areas previously left largely to private enterprise, have transformed the daily life of our religious and threaten to overtax them. The religious life, and in particular the time-table, must be re-thought to correspond to the necessary scale of values and to safeguard the primacy of their religious and apostolic activities.

Social progress, welcome it though we must, has its price. Our religious, enclosed in a network of legislation becoming more and more complex and making more and more demands upon them, facing the falling-off in the vocations which would bring them relief and reinforcement, see their professional duties becoming daily more institutionalized, thereby narrowing the margin of security indispensable to their personal religious development and their apostolic development.

Formerly when a foundress chose education she saw it as a fundamentally religious task; diplomas and technical demands appeared only later. The progress made, which is incontestable, is not without its dangers from the religious and apostolic viewpoints.

Such, in brief, are some of the major causes of a complex

situation in which a variety of circumstances has played a decisive part. We have tried to list them and analyse them, without however being exhaustive, in order to allow us in the second part of this book (intended to outline the programme for reform) to indicate the ways and means towards the necessary solution which, thank God, is within the grasp of our joint efforts.

Part II

TOWARDS AN APOSTOLIC RENEWAL
OF THE RELIGIOUS LIFE

6

THE INNER MEANING OF THE RELIGIOUS VOCATION

WE have come now to the positive and constructive part of this book. We must place ourselves in an attitude of humble welcome before the Lord and ask just one question: What are the thoughts and wishes of the Master on the part to be played by religious in our times? What does He expect from these most precious collaborators? What help does He require from them for the salvation of the world? In reading the following pages we must say with St. Paul on the road to Damascus, "Lord, what wilt thou have me do?" Everything is in that question. How to make God better known, better loved and better served. As Pascal said, only God can tell us about God. Only God can speak about the salvation of souls, whose worth He alone knows, having ransomed them with His blood. He alone is the Master in command, giving the necessary graces in overflowing measure for us to answer His call courageously and to overcome all obstacles. *Da quod jubes et jube quod vis*, said St. Augustine—give what you order and order what you wish. What matters is the glory of God and the salvation of the world: everything else is relative to this absolute. Being open to the breath of the Spirit, to His views not to our own, this is the first thing required of those who, whether from within or from without, are responsible for the destiny of the religious life. For others the sure way is to respond with their whole heart to the demands made of

them by their Superiors. The generous carrying out of
allotted tasks is an infallible channel for the grace which
not only renders their work fruitful but also advances the
time of the hoped-for apostolic revival.

CONGREGATIONS AND SECULAR INSTITUTES

At the threshold of this second part which is intended to
help hasten the 'new spiritual spring-time' for which the
Pope hopes as a result of the Council, it seems important
to make clear the necessity of keeping our traditional
congregations.

As we have said, one frequently hears it repeated in different circles that the classic type of religious has outlived
her *raison d'être* and that the future lies with secular institutes or with some new form altogether. The 'classic'
religious, they say, are prisoners of the past and fettered
by canon law; they can never become the salt of the earth
as one would have wished. Their Rule itself prevents them,
and is an effective obstacle to their penetrating the world.
From this, people conclude that the answer is to let the
past be past, that it is a waste of effort to attempt to move
the mountains of tradition, and better to start from scratch
and make something to measure to fit the needs of our
time.

A curious thing is that this defeatist attitude finds in-
voluntary support among certain religious who think that
apostolic adaptation to the world is the job of secular institutes and that nuns of the traditional type should not
be asked to do more than they do now.

What we want to explain here is why, in spite of the
existence of these secular institutes, the active traditional
congregations are still indispensable and why we firmly
believe in the apostolic wealth which they hold for us
today.

The religious state is defined in the Church today as the

"regular way of living in community in which the faithful agree to obey not only a common rule but also the evangelical counsels by taking vows of poverty, chastity and obedience".

What typifies the regular type of religious life, then, is the life in the community in which the religious consecrates herself to God by the vows of religion. This communal life under obedience, lived within a framework and according to a Rule approved by the Church, constitutes a public affirmation of the transcendency of God and the reality of the supernatural. Religious are witnesses to God's right to be loved and served above all things. They set the world a problem: Whence comes the secret of such vocations? How is this devotion nourished? What is the source of their joy in the midst of the great human suffering to which they minister? Where can one find the key to such a life? What Bergson said of saints and heroes is particularly applicable to them—their existence is a summons.

This in itself demands that they bear a sign, wear a distinctive habit, so that they are recognizable in ordinary circumstances. The habit must be simple and suitable—we shall come back to this—but it must also retain its function as a visible sign. The religious community as such is called upon to bear witness in a manner visible to all.

The secular institute has a different object. In the Constitution *Provida Mater Ecclesia* the Pope defines it as a society "whose members bind themselves to practise the evangelical counsels while still remaining in the world in order to acquire Christian perfection and to exercise their apostolate more fruitfully". This apostolate is exercised by them in their professional life, generally speaking on their own, wearing no distinctive habit, and living as far as possible the normal life of the world.

The comparison of the two kinds of life is alone enough to mark their differences and to show that the religious life answers a different purpose and retains in full its *raison d'être*.

THE PRIMITIVE CHURCH AS PROTOTYPE

The religious community as such constitutes a 'sign' by which the Master ought to be able to make Himself known; an apologetical sign of brotherly love lived according to Jesus' words, "So they may be made perfectly one. So let the world know that it is thou who hast sent me." It is a fraternal community trying to continue in the world the prototype of the primitive Church described in the Acts of the Apostles:

> There was one heart and one soul in all the company of believers; none of them called any of his possessions his own, everything was shared in common. Great was the power with which the apostles testified to the resurrection of our Lord Jesus Christ, and great was the grace that rested on them all. None of them was destitute; all those who owned farms or houses used to sell them, and bring the price of what they had sold to lay it at the apostles' feet, so that each could have what share of it he needed. (Acts IV, 32–35.)

The religious community is linked by its very origin to the *vita apostolica*, to a life lived in the manner of the Apostles. It professes to be the incarnation of the Gospel in all its reality, and therefore particularly in the matter of its social repercussions. That fraternity of Christ's disciples went as far as a certain pooling of property which, although recognizing the right to private ownership, controlled its use for the common good. Religious life continues to express before the eyes of the world the social implications of evangelical fraternity as a consequence of the voluntary association of members and the practice of the three vows.

As Fr. Carpentier writes:

Limited to groups of volunteers, the *vita apostolica* organizes here below a social order and a way of life in which love solves all the problems of mutual understanding. . . . The religious life is seen as a public profession and has therefore a duty and assumes a mandate of 'public' interest, that is of interest to the Church as the supreme universal society. Its mandate is first of all to exist as a community and to bear witness among Christians and in human society. . . . It is also the accredited witness of the perfect evangelical life . . . the permanent charismatic witness to the Gospel's social message to humanity. . . . The religious life is no more than the baptismal life fully evolved along the lines of the social structure of the Gospel.[1]

At a time when communism is trying to impose by force a new social order which destroys the spiritual personality of man, it is more than ever important that the Church should be able to offer the world the picture of living communities, where a voluntary communism reigns, based on divine worship and brotherly love, as a foretaste of what in many aspects a society open to social Christianity and faithful to the Gospel would be.

This apostolic life, it should be noted in passing, includes the supreme reason for its existence—the apostolate *par excellence* which consists of 'testifying with great power to the resurrection of Our Lord Jesus Christ'. This testifying must continue in the Church to the end of time, for the world can only survive on a living faith in the resurrection of Christ.

This, then, is the ultimate mission of our religious communities, and one can see why the Church shows special favour for them and desires to see them flourish.

[1] *L'Episcopat et l'Eglise Universelle*, pp. 427, 430, 433. Le Cerf, Paris, 1962.

All vocations have a reason for their existence, and the Church gives her blessing to them all. It is not our concern here to show the magnificent part which souls dedicated to God in the world can and do play; it is simply to make the point that the existence of secular institutes does not dispense the regular religious from their mission within the world.

THE IDEAL OF THE RELIGIOUS LIFE ALWAYS MODERN

The religious dedicates herself to God—but also to her brethren. To deprive her of the expression of the latter is to misunderstand the nature of her vocation.

St. Benedict used to ask those who sought admission to the novitiate, "What are you looking for?" If one put the same question to a young postulant to some apostolic order or congregation, she would answer, "I want to consecrate my life to God for the salvation of the world, nothing more and nothing less". She has in her, in the magnificent phrase of Gregory XV about St. Ignatius, 'a soul greater than the world'—*animam gerns mundo majorem*. In order the better to consecrate herself to God and souls, she is prepared to renounce the greatest joys of human love, of creating a home. She wishes God alone to be the breath of her soul and the inspiration of her heart.

This humanity which she wishes to serve embraces all souls whom she can reach, especially those who do not yet know Our Saviour and are wandering far from Him. In order to realize this ideal, she accepts the religious vows which are, with reason, represented to her as the great evangelical means of increasing in her the theological virtue of charity in relation to both God and man.

Joyfully she surrenders her heart and her body in order to belong to Christ alone and to have no other spouse.

Joyfully she surrenders her freedom for obedience;

obedience which is not a shackle, but a means of assuring the maximum return from her life and of strengthening and sustaining her apostolic impulse and co-ordinating it more effectively with the work of others.

Joyfully she renounces ownership of goods in order to intensify her absolute availability, in order to be quit of all cares but the service of God.

The vows, as she sees them in this first glimpse, do not appear as bonds associated with the apostolate but as a deliverance and a means to freedom. They are the price of a freedom which places itself at the disposal of the Spirit, which bloweth where it listeth and which is always going forward: "Where the Spirit is, there is liberty."

Such is the ideal glimpsed by the postulant: she must not be disappointed. In sanctioning a congregation the Church gives her guarantee that the way is sure and that faithful adherence to the Rule should lead to spiritual expansion and greater love for others.

Nowadays a postulant will already have had a glimpse of the dedication to the apostolate of which she dreams through her participation in various works and movements before entering the convent. She will already have experienced the joy of revealing Christ to others; will have come across ardent souls; known homes 'in a state of grace, and of mission'—her own perhaps; seen with her own eyes what a Christian presence can achieve for good in the midst of the day's work; and also what a few hours a week devoted to directly apostolic work outside her professional activities can do. She will dream of giving to Our Lord and her fellow-men twenty-four hours out of the twenty-four. She wants her life to be even more unified. One love inspires her, a double love. She wants to go from God to God in prayer, and from God Himself to God in her fellow-creatures in the apostolate.

If her vocation is to overseas missions, the concept of

total dedication is perhaps more striking; but genuine apostolic zeal is a part of all religious vocations.

She renounces everything in order to dedicate herself to spiritual motherhood, to a supernatural fertility incomparably superior to any other, however noble. She wishes to follow the Master wherever He may choose to go.

When a religious chooses an active order rather than a contemplative one (to which other laws apply), it is because she wants to exercise her apostolic zeal in a visible and tangible manner, to contribute not a marginal comment but a part of the very text of life.

SCHOOL OF SANCTITY

The religious life remains for the Church one of the treasures which give expression to her holiness—"I believe in the Holy Catholic Church," we say in the Creed. Religious offer themselves to the Church so that she may lead them to holiness and to nothing less. They enter into the mystery of sanctification of which Our Lord was speaking when He said, "I sanctify myself for them". What in Him was overflowing fullness is for them a progression and an increase. With Him and in Him they hope to sanctify themselves for the world. By renouncing human love they sanctify the conjugal love of others; by their renunciation of riches they are an invitation to detachment from the material possessions that so often dog mankind; by their renunciation of their own will they proclaim the fact that adherence to the will of God is the only liberty.

They sanctify themselves for the world.

They sanctify themselves also by sanctifying the world. If it is true that no one can give what he does not possess, it is equally true that the surest way to have anything is to give it away: spiritual treasure can only be possessed when one passes it on to others; one keeps only what one loses. In the words of Christ, "It is the man who loses his life for my sake that will save it".

Magnificent ebb and flow: sanctity increasing *in order* to give more of itself and increasing *because* it gives itself.

BELONGING TO GOD

Striving for sanctity for and through the apostolate, the religious belongs to Our Lord's chosen and is the part reserved for Himself. She embodies in her own way what our Blessed Lady was for God, in the words of St. Louis Marie de Montfort: "Mary is God's paradise and His one chaste world where the Son of God entered in order to work His wonders, to dwell there and enjoy it."

The spouse of Christ, the religious, is the result of the triumph of His grace: she is the beam of victory of Jesus who in every age finds some souls who leave everything for Him. In answering His summons, the religious quits a world which she knew and in which the only thing she despised was sin. She renounces for a higher love treasures whose value she understands perfectly well; yet she joyfully offers Our Lord all that makes the world so attractive. It is a triumph for the primacy of God.

BELONGING TO THE WORLD

Nevertheless, the religious brings with her the world she leaves behind and finds it in a supernatural way in the very heart of her vocation.

When God gives a soul that grace of choice which is what a vocation is, He is undoubtedly showing a special love, but this love embraces all mankind. Loved by God for her own sake, the religious is also seen by Him as the representative of her race.

When God loves, or chooses, or shows a preference, He does not do so as we do. When we prefer one thing to others, or choose one thing from among others, what we do not prefer or choose necessarily falls back to second place. But when God loves or favours it is in order that the whole world may thereby be enriched. She whom God loved above others, whom He chose from among all creatures, who was 'blessed among women', was loved and chosen and blessed in order to become the Mother of Christ and thereby the mother of all mankind. Our tendencies and our capabilities are made to match the particular form of apostolate that God asks of us. It is through us that God intends to love others; it follows, then, that each grace we have received must be converted into apostolic action for the benefit of the world.

The religious gives herself to God, but also brings with her in her striving towards God the whole of mankind for whom she makes herself responsible. For her more than for anybody should St. Augustine's words be true: "Do you want to know if God is there? When you turn to Him, have you the interest of humanity in your heart? When you try to approach God, do you bring with you human kind and all its cares? Do you bring with you those whom He gave you to love? If mankind is present in your tenderness and love, God is there."

The religious is not of the world, but she is in it; she takes it with her when she enters her convent; and the world expects the enrichment it has a right to as the result of her sacrifice. If the religious is not to fail God she must not fail mankind. She fulfils a function as mediatrix between God and man, and a mediatrix must join and bind together the two banks of the river; she must throw a bridge across and see that it is firmly anchored on both sides. If she belongs fully to her time and is rooted in the world as it is, the religious can only realistically and logically accept the conditions which will allow her to fulfil her proper mission in the Church today.

7

RETURN TO SOURCES

WE must keep this ideal of the religious life constantly before our eyes if we wish to do our part, so far as it concerns us, in making reality coincide with the ideal. We cannot disappoint this young girl who has seen the possibility of giving herself to God and mankind and wishes to do so; the religious life she finds on entering must correspond to her highest aspirations.

In order to give the religious vocation its proper value in the world today, it is essential to establish quite clearly the balance between the religious life and the apostolic life. So long as there is dualism there will be conflict. There can be no question of opting for the religious life at the expense of the apostolate, or for the apostolate to the detriment of the religious life. The religious life and the apostolic life are not antonymous but more nearly synonymous, each containing or implying the other. So long as one thinks in terms of defending the religious life against the encroachment of the apostolate, or of encouraging it at the expense of the apostolate, one's approach is superficial and no solution is possible. The demands of the apostolate grow out of the religious demands as a flower stems from its parent plant. The only reason for consecrating one's life to God for the salvation of the world is to achieve a greater apostolic radiancy than one could in the world. We believe that this fundamental unity and harmony will best be seen if we go back to the origins of the religious vocation as they are to be seen in the Gospels and the Acts of the Apostles, as they emerge in the course of history and are exemplified in the lives of the Foundresses. This return to tradition—and, if need be, beyond tradition—allows us to recapture Péguy's 'first morning' and to relive in their pristine freshness the intuitions that led to the birth of religious congregations.

This return to the well-springs leads us first of all, of course, to what is and always will be the model for all Christian life and all religious life—the Gospel. It is here that we must seek the thoughts and wishes of the Master.

THE GOSPEL

No constitutions or customs can override the Gospel. The value of any rule derives from its constant reference to the one valid basis, the word of God. Nothing in the religious life may conflict with a single verse of the Gospel, for it exists only to translate each line of it into the language of everyday life and to give it material existence. All that one should be able to read between the lines of any constitution is the Gospel which should be its very fabric.

Now the Gospel is not a book based on fear and timidity. On every page we are invited to go forward boldly in the name of the Lord into the world that is to be saved. It is not only by words and phrases that this invitation is conveyed, but also by the fundamental mystery the Gospel reveals to us—the Incarnation of the Son of God.

The Word, Scripture tells us, thought it no deprivation to leave the bosom of His Father to come into this world to redeem it by His Incarnation, by becoming one of us, so that we might partake of the divine. This prodigious fact, which is emblazoned across our horizon like a rainbow, constitutes the most astonishing appeal to us to 'approach', to make little of distances, to cross the abysses and to be all things to all men.

THE MASTER'S EXAMPLE

The example of the Master who at one bound—*exultavit ut gigas*—passed from Heaven to earth to 'save that which was lost' is our eternal model.

Our Lord came and He went; it is up to us, too, to go in search of that which was lost and must be saved. Jesus came and multiplied His contacts with men in order to reveal His Father to them. Christianity, let us not forget, is first and foremost the bringing of God to man, the communication of His truth and His life. The Church is Jesus Christ 'published and broadcast'.

The Master came and He went to His brothers by adoption. These included His disciples, naturally, but also the Pharisees and the Sadducees, the Samaritans whom one 'simply doesn't know', and the sinners whom one avoids.

He came and He went as a doctor goes to the sick and not to the healthy. He said so Himself.

He came and He moved in circles which 'respectable' people despised and thought it unhealthy to frequent.

He came and He went into the house of Zaccheus and walked in the market place and on the roads and paths of His country, unceasingly, tirelessly.

He came and He went among men and spoke with them.

He spoke sometimes cryptically, sometimes clearly, according to the circumstances. Sometimes he spoke but one word, or a half-finished phrase, or he made a long speech, as on the Mount; sometimes it was a point-blank question, sometimes the answer to another's question, sometimes a meaningful silence. But He spoke to men, whether assembled in crowds or encountered individually at the village well or on the shores of the lake. He spoke as 'no other man has spoken' and His theme was always the same : He spoke about His Father so that men might discover that they were the sons of God and brothers of one another.

He spoke because He was the Ambassador of God, the prototype of all missionaries.

He spoke because He was, in every fibre of His being, the Living Word, the Word made flesh.

From this we can realize that a Christian is not a Christian if he does not participate in this same mission, if he does not echo this same Word, if he does not in his turn bring God to man with all his strength and energy.

THE MASTER'S COMMAND

To the example of His own life Jesus added the explicit command to go into the world in His name.

One command is written in words of fire in the Gospel: to go out into the world and bring the Gospel to all men.

It is an order that admits of no 'attenuation by explanation', no exegetical restrictions, no plea of impossibility.[1]

It is the order that governs every life dedicated to God and our fellow-men: in that life we must keep it ever before our eyes.

It was the final charge given the Apostles at the end of three years of instruction during which He had never ceased to make it clear.

"As the Father sent me, so I also send you."

"I send you as sheep among wolves."

"I send you to the lost sheep of Israel."

The Gospel is full of such exhortations in which the Master charges His disciples to be the heralds who will go out into the highways and byways to invite the many to God's banquet.

Everything in the Gospel breathes calm courage and the certainty of a divine assistance that will never be lacking. "You will do greater things than I."

St. Paul was the faithful echo of Our Lord when he wrote

[1] Cf. *L'Eglise en état de mission*, Ch. VIII, Le Commandement du Seigneur, where this theme is developed.

to Timothy words which should figure in the heading of all Rules of Life: "The Spirit he has bestowed on us is not one that shrinks from danger; it is a spirit of action, of love and of discipline."

THE ACTS OF THE APOSTLES

The whole Gospel is an invitation to the direct apostolate, to walk upon the waters.

And this is how the disciples understood it. The Acts of the Apostles are just an account of the courageous loyalty of Jesus' disciples at odds with the pagan world they were trying to penetrate. "We cannot not speak," they were to tell their judges. Among these disciples, moreover, women had a special place.

It was women who, disregarding the danger of compromising themselves, accompanied Jesus on the road to Calvary while the Apostles, with the exception of John, had fled.

They appear again, loyal and unbroken, at the foot of the Cross, at the empty tomb on Easter morning.

Women gave themselves fearlessly to the propagation of the faith and the service of the infant Church even though for some of them it meant martyrdom.

They are constantly met, either as deaconesses or as private persons, in every branch of the apostolate. It is worth reading again certain passages in the Acts, or in St. Paul's epistles, where he lists the intrepid women who worked with him. Nowhere in Scripture is there any trace of contempt or disdain for women. Jesus trusted them— and they did not disappoint His trust. If we really want to go back to sources, reading the Acts is an essential item on any programme of apostolic reform in any field.

THE FOUNDRESSES

The Acts of the Apostles are carried on throughout the

history of the Church, in particular by those generous and saintly souls who were responsible for the founding of the various religious orders and congregations. Their lives are a faithful echo of the Gospel.

To bring out the missionary zeal that underlay the foundation of the active congregations, nothing is better than to study the life of the Foundresses.

H.H. Pope Pius XII emphasized this point very strongly in the repeated and moving appeals he made for them to be brought up to date.

"Most of the time," he said during the Congress of Religious in Rome, "those who wrote the rules of religious institutes planned their new foundations in order to fulfil some urgent function or to answer a need that suddenly appeared in the Church and had to be dealt with then and there. If, therefore, you want to follow the example of your Foundresses, your attitude must be their attitude."

Many of these Foundresses have been canonized, which means that the Church has formally set the seal of her approval on their life and purpose. What they all had in common—and some of them to an almost unbelievable degree—was the apostolic zeal which led them to minister to material and spiritual distress. They went further and further in their efforts to save souls and to make Christ known to all men everywhere.

Their history is glorious—and at the same time sad: glorious on account of the Foundresses who represent initiative, tenacity and Christian audacity; sad on account of the opposition they encountered from men who were slaves of the anti-feminist prejudices of their time and prisoners of routine or of a too narrow canonical definition of the term 'religious'.

The history of congregations is all too often the account of a series of obstacles to be surmounted in order to win the right to go where the distress to be relieved was most pressing, of the right to the freedom of action vital to their work.

In so many cases the story of the Foundresses recalls that of Veronica, braving the criticism and hostility of the mob to come to the Master to wipe the blood and dirt from His face. Their zeal was directed to Christ crucified in the souls of sinners, to Christ unknown or wrongly known. Like Veronica, they were not concerned to know whether it was ladylike to mingle with a crowd of Pharisees and soldiers; they went to Christ with the courage born of love —and God blessed their action and their work.

Another striking thing is that the Foundresses never lost sight of the fact that their *raison d'être* and that of the infant foundation was to bring Christ to those in material or spiritual distress. They faced sin and sought the sinner, the lost sheep. They took Our Lord at His word and went out to invite people to come and be nourished at the eucharistic banquet and take sustenance from the word of God. They went. . . . One might almost use that as a sort of refrain in their lives which were in no way enclosed or cloistered. They did not wait within walls until the police brought abandoned infants or delinquents to the door; they were imbued with the mystery of the visitation and hurried with Our Lady into the hill country to be of service to others.

The sick, delinquents and children were their special care, both for their material needs and for their religious welfare.

The work of the first companions who rallied to them was, by human standards, on a very modest scale, but they never lost sight of the directly religious aim they had adopted.

After the death of the Foundress the work grew. Where to start with there had been a bare handful of sick or orphans, their numbers today are reckoned in hundreds or thousands. Socially this undoubtedly represents remark-able progress, but a progress which from the apostolic viewpoint has been dearly paid for. Technical progress, and mere weight of numbers on top of that, have obscured the original insight of the Foundresses whose first care was the work of spreading the Gospel. We must therefore con-stantly come back to first origins if we are to maintain a balance between the various components of the religious life.

How this will work out, we shall have to see as we go along. Suffice it to mention here the extent to which the lives of the Foundresses show up the main points and pre-vent the wood from becoming obscured by the trees. To recapture the initial inspiration is to breathe the sweet air of the Gospel. Times have changed and methods have be-come outdated, but we must not betray the original spirit. All communities must guard jealously, or rediscover, the original 'marching orders' which made their Foun-dresses pioneers of the direct apostolate. On the subject of true loyalty to tradition, the following words of J. Chevalier are very true :

> An identity of life supposes a continuous change whose very continuity assures unity. In time there is always some element which is changing, but it only changes in order that some other element may remain constant.[1]

Let us, then, examine the main activities of religious to-day in the light of their origins, and try to see what avail-able or latent apostolic possibilities we can find there and how they can best be enlarged.

[1] Quoted in *Equilibre et Apostolat*, p. 238.

8

THE WIDER VIEW

Deployment for the Apostolate

BEFORE going exactly into what constitutes for the main categories of religious the apostolic advancement which is the theme of this book, we should like to draw attention to some fields of action which have not been opened up or which lie fallow in part and which should be shared among these categories according to their particular gifts. One can, and indeed must, admire what is already being done, but obsession with the salvation of the world urges us to find out what still remains to be done and makes us worry about the harvest abandoned for lack of labourers. To the man who said he had faithfully fulfilled the law, Jesus said, "In one thing thou art still wanting . . ." In the spirit of the Gospel we should try to find out what duties we have not yet fulfilled, rather than enumerate those which we have done our best to fulfil. It is not a question of abandoning the duties of one's state, but of seeing just how far these duties can be extended; it is not a question of embarking on new tasks instead of traditional ones, but of carrying the latter to completion, of following one's vocation to its end, of really being what one sets out to be.

This extension which is indispensable for apostolic reasons is also essential to the balanced development of our nuns themselves and to the survival of our schools and institutions.

Growing State control in the fields of education and medicine put the accent on the professional side of the life of religious. Officialism threatens them: both teachers and nurses sometimes feel themselves to be more teacher or nurse than nun. This is the cause of the unease of which we have already spoken.

"One thing is still wanting . . ." This one thing, though often not clearly seen or not seen at all, is the possibility of a more direct apostolate.

It is not a question of adding new tasks: the nun's day, like that of the rest of us, has only twenty-four hours. It is a matter of the practical revision of the scale of values and of reserving a special place for certain specified apostolic activities in the same way as a special place is held by the spiritual exercises prescribed by the Rule. Only if this is done can the real professional life of nuns be fruitful. Their spiritual and apostolic activities are the special portion that God reserves to Himself.

We do not intend to put forth a fixed and invariable definition of what 'God's portion' will consist of. It must be made to suit each case and should be introduced gradually in a manner still to be decided upon. In the next chapter we indicate some adaptations which may make the change-over easier. The essential is to realize that this integration is indispensable for the full vitality of a soul vowed to God, for, let us not forget, in choosing the religious life the postulant was not principally drawn by the prospect of becoming a teacher or a hospital worker: she wanted these activities to be the means whereby she played her part in the salvation of the world. It is in any case only to the extent that she is clearly fully apostolic that a nun will draw others to follow in her footsteps. A religious must be in a position to convey the best of herself.

It is essential to her spiritual joy that she be able to communicate to others the secret of her great love. It is hard to imagine a girl who could never talk to anyone about her fiancé: she needs to make him known and appreciated by others. "I believed, therefore I have spoken," said St. Paul. From the overflow of his faith words sprang up— they had to. Silence is agony for anyone who wishes to proclaim his discovery of the Messiah from the housetops. The religious, the bride of Christ, knows the same feeling and demands to be given the means of pursuing her apostolic vocation to the end. Her task is not done until she can give Christ to others by gathering round her other souls whom she can fire with love of God, for whom she can make Christ live in every aspect of life, individual, in the family or in society; whom she can teach to proclaim Him by their own lives in their respective circles. A whole field of apostolic action among the laity opens up for the nun who knows how to discover such people, how to group them into appropriate Catholic Action movements already existing, or into some new movement to meet a present need, how to inspire their goodwill and keep it alive.

It is an immense field with boundless perspectives, this apostolate of bringing Christ to those who do not know Him, and introducing Christian standards into every aspect of the lives of so many who are Christian in name alone. The world is full of religious illiterates who cannot read God in the book of life, as it is full of Christians who have done no more than glance through a few pages, if that, of the Gospel. This is the world that claims the nun's attention.

Certainly not every religious is suited to every kind of apostolic action; the particular kind of activity must be chosen to suit each individual, including the Sister-cook, and an apostolic outlet must be found for each in keeping with her capabilities. We must be firmly convinced that, in comparison with the surrounding world which is spiritually undernourished to a scarcely credible degree, our nuns are overnourished and have a surplus they can distribute to those around them. Consider the time devoted to the nourishment of their souls: Mass and Holy Communion, meditation, spiritual reading, instruction, recollection, retreats—all these are storing up divine energy in them, making as it were high-tension storage batteries of them. Consider also how the Church surrounds them with loving care: the Superiors' orders, the Rule, the customs of the house, examination of conscience, spiritual direction, mutual support, canonical visitors, and so on all the way up to the Sacred Congregation for the Affairs of Religious— everything combines to ensure their spiritual welfare and to fit them ideally to be the inspiration of the lay apostolate among women. It would not be reasonable if all this care and trouble ended in their lives being concentrated on a restricted number of pupils or patients without there being also some outlet to the world in need of salvation. It would not be reasonable for the life of a religious to be turned back on itself and sheltered from the winds blowing in the world outside. Nuns do not dedicate themselves to God in order to live a petty, patterned life in a confined space cut off from the world.

It remains for us now to explain the extent and nature of the apostolic element to be incorporated in the religious life.

Today nuns are particularly active in education and the care of the sick, and it is in these two fields that they are most numerous. It is there, too, that we must go to study the question of apostolic yield. What we find for them can easily be applied to other categories by making the necessary minor adjustments.

EDUCATION

The Basic Apostolate

To avoid all possibility of misunderstanding, we must

start by saying that the educational work carried out by our religious institutes is in itself an apostolate.

We must repeat here all that we said in our book on the education question, about the climate of Christianity in which the instruction is given and which is so necessary for baptized children. The Christian outlook on life is not conveyed to the child by religious instruction alone: it is in all subjects, and particularly those which concern man more closely—history, literature, ethnology—that youth learns to see everything in the light of Christ and to acquire that integrity of judgment without which the Christian life is impossible.

The forming of youth is in itself an apostolic work so long as it respects the essential aim. Why, in the last analysis, do we develop the child's faculties by teaching him algebra or Latin or history if it is not to increase the value of this growing Christian, to make him better suited tomorrow to spread the thought and the life of Christ in the world which will be his? This *consecratio mundi*, this christianization of the world, can be achieved only by fully human and fully valid Christians. Only to the extent that Christianity has penetrated them will they be able to radiate it to other individuals and to society.

There is no such thing as the 'natural order'; supernatural order is the only real order, willed by God, and it is in this order that all human talents are deployed. When he undertakes to make a man of a Christian, what the educator is undertaking is to turn him into a complete Christian. This can only be achieved by the continuity and harmony of an integrated education where all disciplines work together towards a common goal.

Each lesson, even the most remote from sacred things, is impregnated with the whole atmosphere of a class or a house.

Each subject taught can and must be integrated in an

overall Christian perspective. There can be no dualism between science and faith, since they are approaches to the same truth that comes from God. Teaching itself is only one element of the living synthesis of true Christian education. To be successful, Christian education demands that it is not only the mind that must be formed but also the will, the character and social and apostolic sense. All this supposes coherence and co-operation among educators; it supposes also an incontestable apostolic effort, which is the very soul of our institutions. The religious, whose whole being is vowed to God, is especially qualified to give the young such an integrated education which, being centred on God, embraces and unifies all the different facets of human culture.

SPECIFIC TASKS

What we have said so far applies to the fundamental apostolate inherent in our institutions in so far as they are logical in their Christianity. They are the stock on to which specific apostolic tasks must be grafted if we wish Christian education to give its maximum yield and fulfil the Church's hopes.

Let us glance now at the various branches of the field entrusted to teaching nuns and see if we cannot find some occasions for apostolic action.

First of all, within the schools.

Apostolate in Schools

It is not enough that nuns themselves are fully conscious of the apostolic duty. They must communicate this conviction to the girls in their care.

Their young charges must be educated to a Christianity that is complete—that is, apostolic. This means that each pupil must receive practical and progressive training. And this must be done in the ordinary way of school life. One must be able to offer the pupils a variety of works and movements capable of stirring their enthusiasms and generosity so that they can do an apprenticeship in social and apostolic work. A single standardized voluntary–compulsory movement cannot take into account the diversity of talent and vocation among the girls. Such movements and organizations as can be made available would gain by being affiliated to diocesan, national or international societies, since they would profit from their know-how and experience. However, certain organizations of a purely local character are not excluded. Too often in our schools there are only a very few girls engaged in any such activity, a tiny minority. We cannot accept this anomalous situation. It must be understood that a fully Christian education must be given to all pupils; the means to achieve this must therefore be provided. Once the principle is agreed and the means at hand, some girls will choose a movement that is directly apostolic while others will go for something which is of a more social or charitable nature, but which indirectly leads to apostolic contacts. There are many mansions in the Father's house—but they should all be inhabited.

As in English schools, where the time-table allows a set number of periods for physical culture but leaves the choice of games or sport to the individual, our institutions must allow for a choice of apostolic or social work. Parents should know that this is part and parcel of Christian education, and it would be a good idea to explain to them what such an education entails.

The conservation of our faith is also at stake. We are no longer in the age of traditional Christianity handed on from one generation to the next. To remain alive, faith must propagate : like fire, it must set alight to its surroundings or it will burn out and die. A generation that has not learned to radiate its faith, to communicate it to others, is doomed to spiritual sterility. Catholic schools were not set up for the purpose of presenting diplomas to young ladies entrusted by their parents to our care, but in order to make out of them apostles with diplomas. A school does not earn the title of Catholic just by giving religious instruction and making it possible for the pupils to attend Mass. For a school to be Christian it must produce committed Christians needed by the Church and the world. This is a perfectly normal requirement of Christianity.

It must not happen, an American bishop once said to us, that our colleges produce nothing but Old Boys or Old Girls. Our education is successful only if it turns out a dynamic, conquering, Christian faith. Anything else is failure.

It might sometimes happen that the very number of pupils makes it impossible to organize things so as to ensure this kind of fully Christian education. In that case we shall have to be bold and cut down the numbers, for if we do not, we shall be defeating our own ends, which are not to get as many diplomas or scholarships as possible but to turn out girls capable of playing the part that Our Lord expects of them in society.

After-school Apostolate

If education is to be an integral affair, teachers cannot just forget about their pupils when they leave school. The whole complex of the pupils' lives should be the object of their active and fruitful solicitude.

Formerly the pupils confided to the nuns' care lived like them in a closed world: society was Christian or at least imbued with Christian morality. But this traditional Christianity is giving way to Christianity as a matter of personal preference. Today when a child leaves school it has often to go to a family profoundly unsettled on both human and Christian planes, and in any case to a world becoming more and more pagan, in which everything is a matter for doubt and questioning.

Life after school threatens to undo what education has

achieved *intra muros*. This is why nuns must take an interest in it. But this negative reason is not the sole reason: education must be a co-operative effort by teachers and parents. Upon this depends the efficacy of an action that should be continuous and complementary. A child's education is at the mercy of the parents who can either further it or undo it altogether: their co-operation is essential. It is the job of the religious to make them understand this, and she should be trained up to a certain point to undertake the Christian education of parents, for one cannot separate what life has bound together with the closest ties. Education is one: school and family are like the two halves of a gothic arch. The education of children, as one philosopher expressed it, is the family reaching maturity. The religious has therefore a part to play in ensuring the continuity of this common work.

A teaching nun wrote to us very sensibly on the subject of this continuity:

Experience has shown that all is not done when one has spent oneself without counting the cost to provide our children with a Christian education. This education is the foundation of Christian society, but we must also help in the construction of the entire edifice in the persons of the young adults, and so continue our work as educators right to the end.

Thinking of the nuns' position, she added:

But what can we do? Our rules are framed in terms of the world as our Foundresses knew it; they did not envisage the world of their successors, a world necessarily very different from theirs. So we see our nuns vowed, if nothing is done to remedy the situation, to be the guardians of only those sheep that are within the fold and to allow to stray those sheep which, once outside the fold with no shepherd to follow them, leave the flock and get

G

lost in such great numbers along the road. Yet how great is the price of a soul! Christ died as much for grown-ups as for the souls of children. We must therefore go out and find them wherever they may be.

Family Apostolate

School is more than just an assembly of pupils. Each child is an open door to a family, a door revealing endless vistas of family action. First there are the natural contacts: parents like to meet their daughters' teachers. These contacts must not be restricted to Reverend Mother or the Directress of Studies but should extend to class mistresses and any who have direct responsibility for the girls.

Parents' associations, which happily are being formed in various places all over the world, can be an increasing occasion for contacts: one must make the most of the latent apostolic opportunities they offer. What a wonderful thing it would be to have a hand in the steady, solid work of forming Christian families and organizing the best of them into various family movements or Catholic Action movements aimed at christianizing the others. But this supposes that one does not sit idly by and wait for the parents to come and ask for an interview. It means that educators must approach the families. Of these families those who do not come of their own accord are often the ones who most need attention. This supposes, too, that the 'family' part of apostolic work is recognized as an integral part of the duties of the educator's state.

It also supposes a freedom of movement which, in turn, demands some necessary adaptations. If the end is desired, the means to it must also be sought: and there can be no doubt about the desirability of an end demanded by the salvation of souls.

It is hard to overstress the importance of this 'missionary' action in the basic cell of the social body, the family. In visiting the parents at home one has the benefit of all that this represents of reality, stability and human truth. At home the child is natural, genuine, her true self—and so are her parents. Outside the home they are subjected to a mass of influences which make them adopt artificial attitudes. This is an elementary observation. Social pressures exert their full force as soon as a man steps outside his own front door; if these pressures do penetrate unbeknown into the home, then at least he is in a better position to resist them there and to exercise his freedom. His true nature is stripped of its layers of social varnish and asserts itself at home. And here, too, a man may make what use he will of his freedom. To bring one's message into the home is to deliver it in an atmosphere of freedom where it stands a far better chance of lasting effect than otherwise.

Lay Staff

Nuns, however, do not act alone. More and more, they share their educational responsibilities with lay mistresses. This fact is of great importance in the matter we are considering. Nuns alone cannot suffice to ensure that each pupil is educated to the apostolate: lay mistresses and lay professors must be brought in, and this means that they too must have training to fit them for this task: they must, as it were, take a degree in 'apostolics'. Who is to blame for the frequently heard complaint that the collaboration of the lay staff is restricted to their professional sphere? Have we prepared them for the apostolic role in the instruction we gave them in teacher training? Have we trained them, initiated them in apostolic methodology as they were initiated in the methodology of languages or the sciences? There is a vicious circle here and it is a matter of urgency that it be broken. Once this is done, instead of regarding the lay invasion as a necessary evil to be borne with, we can see it as a providential opportunity to relieve our nuns of a part of their profane duties and enable them to devote themselves more effectively to a wider apostolate which should include in particular the post-school world—that is, first and foremost, their former pupils.

'Old Girls'

Too often the organization which brings Old Girls of a school together once a year or so is rather academic and has no apostolic or social significance. Nevertheless, these souls remain in the charge of their mistresses whose mission is not completed until a Christian home has been established and everything has been done to make it a success worthy of the children of God. Their mistresses must keep in touch with them, devote some time to them, take an interest in their lives and try to get them interested in some apostolic or social work suitable to the circumstances in which God has placed them. Many factors will doubtless arise to weaken contact between mistresses and girls, not least among them the desire of the latter to be on their own and be 'allowed to grow up'. But it is the mistresses' job to find the right psychological approach that will enable them to maintain or create contacts. By means of existing movements, or new movements to be created for the purpose, somehow the work of education must be carried on among the young grown-ups. For some of them this contact will be, by the grace of God, a determining factor in awakening or sustaining the beginnings of a vocation; how else can one hope to preserve a vocation if all contact is lost at the very time when life's most important decisions are being made?

For the others this contact will be a permanent source of grace, always at hand, according to circumstances. All that we said in *Love and Control* about the part to be played by educators in preparing their charges for bethothal and marriage should be repeated here.

But how can these needs be met if a nun cannot be near her old pupils; if she cannot reach them; if the Rule does not allow her the means to go to them; if her Constitutions forbid such contacts; if in fact this apostolic aspect of her task is not written into the Rule itself; if the Rule does not both fix the proper conditions and ensure the possibility?

Outsiders

Ought we not also to think about the other young people, not Old Girls of our schools, who must also be taken under our care?

Certainly our existing works and youth movements do their best to teach these young women, and we must do what we can to help them in this. But there will always be a number who will have nothing to do with any movement, yet must be evangelized. The field of young women is far too often left uncultivated. Who better than our nuns could take on the job?

It will not always be done directly by personal contact, but perhaps by organizing young Christian women in such a manner that they can penetrate the area. Who has not heard of moving examples of lay apostles radiating a warm and living Christianity in so-called impenetrable areas?

Unfortunately these lay apostles are a mere handful. It is up to religious to find more of them, to group them together and train them and stimulate them.

Clearly this apostolic 'border action' has to be organized. This could be done by several religious congregations working in concert to an overall plan. Why should we not envisage qualified nuns giving courses in religion in State schools? Or running catechism centres? Or Press centres? Nuns have to collaborate with the clergy and with the apostolic movements, in a suitable manner and in line with their particular object, in the evangelization of young grown-up women who live outside the influence of institutions.

The pedagogics of such action have to be worked out; all we are doing here is to draw attention to a world in distress waiting to be christianized by our educators.

The 'Poor' and the 'Abandoned'

Older adolescents are the abandoned children of our era. Formerly a Vincent de Paul or a Jean-Baptiste de la Salle collected from the streets the small children who had been materially, morally or intellectually abandoned. Today young adults—that is, between 16 and 25—are a class particularly in need of attention. They are the spiritual orphans of our time, up against the problems of life, alone, turned in on themselves, yet so anxious to be understood and to be tactfully and discreetly guided. The unpleasant exploits the papers report from everywhere in the world are the work of victims of this abandonment. No doubt the parents bear the primary responsibility for this state of affairs which is too often due to moral poverty in the home—but we are not concerned here with analysing causes but with finding remedies.

Following their Foundresses in their love for children in distress, the nuns of today will transfer their love to the abandoned adolescent, convinced that she has more than ever the need to be loved and understood and helped.

Given the choice of means, one would like to open schools which would enable an active principle to work among these girls. The same pastoral concern would dictate the choice, whenever possible, of the sort of school which prepares girls of different social backgrounds for occupations that are rich in human contacts.

Formerly the Foundresses showed a particular love for the poor. Free schooling means that as far as education is concerned there are no longer any poor. But "the poor we have always with us". The evangelical preferences of the Foundresses would be best respected today by opening clubs or other places where young women deprived of normal Christian influence will be welcomed and can spend their free time. Modern society accords an ever-increasing place in life to leisure and amusement, and it would be hard to overestimate the importance that the environment of leisure has assumed. All mass communication media are propagating a concept of life which is invading our homes and capturing the adolescents; and it is a concept that is materialistic, pagan and as far removed from Christianity as it is possible to be.

The teaching nun is the obvious person to teach girls how to organize and occupy their spare time in a profitable manner—avoiding the dangers.

The environment of leisure extends beyond the home: travel, week-ends and holiday camps have assumed great importance in the lives of our young people. The presence of a nun during these times of naturalness and openness would be a great blessing for them.

We asked a group of Mothers General how they would answer the classic objection which tends to reject any new ideas and which is expressed something like this: "The duties of our state as teachers already constitute an active apostolate. Why look for anything else?" Their unanimous reply, which seems to us to be clear, courageous and conclusive, was:

"Teaching is sometimes the only way for the Church to get a foothold in, say, mission lands.

"But this teaching, even of non-religious subjects, must always quite definitely be done in the service of evangelization, religious education and the direct apostolate.

"Under pressure of circumstances, burdened with the preparation of new courses, the correction of exercises and a multitude of disciplinary and administrative chores, a nun no longer has time for the essential; she risks becoming a mere distributor of texts, an invigilator, with practically no real apostolic contact with her pupils.

"There is no question of denying the supernatural merit of these tasks carried out under obedience and often with love; but they no longer have any value as witness. Sometimes the only impression they convey is of being all too human—or not human enough—and are a source of scandal. This is obviously not what our Foundresses intended.

"This deviation from the original intention is resented by those with vocations. Particularly if one has been active and had a certain amount of responsibility in apostolic movements, one wants to give one's life, not to become a supervisor or a grammarian but to bring Christ to souls. Neither the pupils nor the lay staff have sufficient opportunities of seeing the nuns in their true role of persons dedicated to love and the service of Our Lord. And what might not be said of the spiritual crises and difficulties in the spiritual development of some of our religious! We should like to make our own the words of Mgr. Maury addressed to priests in charge of religious: 'Whatever position a nun hold in the Church, she must have the task of evangelization and the opportunity to carry it out.' "

CHARITY

Charitable Works

Another field in which our religious expend themselves with so great generosity is that which comprises hospital and social work, mainly for the benefit of the sick, the handicapped and the aged. Nuns are to be found everywhere where suffering humanity calls for help: to them is confided the ministry of the Church's maternal pity. Their devotion moves people and wins their hearts. This living year-long example of charity forces respect.

The sick expect to be cared for physically, no doubt, but they expect something else from nuns—moral comfort, support in their trials, attentiveness. Perhaps attention is what the sick need most: they want someone to listen to their life-history and their troubles, and it is by listening patiently for hours that the religious wins the right to speak and to utter the few words which count. She has the delicate task of gradually introducing to her patients the mystery of suffering, to show them how God can be met in the midst of pain and how this encounter can put everything in its proper perspective. In order to fulfil this mission, which is inherent in her vocation as a nursing nun, she must have the time, the secret of personal contact and the ability to communicate.

We have seen that the increasing complications in the organization of nursing tend to limit the nun to purely technical tasks, perhaps as ward sister or supervising the work of lay nurses and probationers. She is so overworked that the most she can hope for is occasional fleeting contact with the sick.

Devotion and Evangelization

There is no doubt that the religious consecration of the nun gives her work a supernatural value, but it still remains to complete this work by the sort of directly apostolic work we envisage for teaching nuns. This apostolic complement to her technical work is vital not only to development of the nun herself but also to foster vocations. The danger of officialism is perhaps even greater for nursing nuns, for medical care does not produce the same directly apostolic opportunities as teaching does.

Nuns know and feel that devotion and evangelization are not synonymous. Their unselfish devotion to the sick rouses everyone's admiration and often paves the way for more intimate contact, but devotion as such is not the apostolate.

This does not mean that devotion must be reduced: only that some means must be found to complete it by direct apostolic action. Our Lord's share must be kept sacred, both for the sake of the sick and for that of the religious herself, who is thereby enabled to give the best of herself. Devotion, which is giving oneself, prepares her for the apostolate, which is the giving of Another through herself. If this is kept in mind we shall begin to see all sorts of valuable but previously hidden possibilities for spiritual radiancy. Many of these possibilities can become fact provided that a certain amount of time is made available to fit them into the routine of life.

Steps to be Taken

There are first of all certain steps to be taken with regard to the sick collectively or individually. Individual contacts with them of a less fleeting and superficial nature will encourage their confidence. A great many moral troubles are confided to nuns who come to the sick with Christ in their hearts. There are lesions undisclosed by X-ray which need her treatment; there is suffering which longs albeit unconsciously to know its Saviour. Among collective approaches we have communication media like broadcasting where a 'Sick People's Hour' in appropriate circumstances could create a tonic atmosphere that would benefit both body and soul.

A sick person is normally not alone in the world. To contact with him, should be added contact with his family. In some cases this will reveal a need for social services; or in the intimacy of the home it may become apparent that there are religious deficiencies, situations that need to be set to rights, or sin at work. Briefly, a vast field for apostolic action can be opened up by anyone who has eyes to see and ears to hear and the gift of gaining people's confidence. In due course patients will return to the homes that the religious has in the meantime got to know; it should be possible to follow them there, or to have someone do so, to assure oneself that they are keeping the good resolutions they have made. Suffering is like a retreat during which Our Lord speaks without words to the heart of the sick person in his weakness and dependence on others.

In the same line of thought, why should we not try to organize days of recollection or retreats for ex-patients—and their families? Or if it cannot be done on the spot, a little discreet propaganda for existing retreat houses would be a valuable contribution. The dialogue which physical suffering so often starts between God and a soul must be followed up and its fruits be brought into everyday family life. In addition to what one can do oneself, there is a vast field in which one can get things done by others if one knows how to organize and train them.

Action by religious to increase their co-workers in the para-medical field concerned with nursing must be undertaken. First of all this must be done in the nursing schools, which must be treated as 'mission territory' just as other schools, and later among probationers, lay staff and others within the orbit of our hospitals.

Although it is true that the hospital nun is in greater danger of becoming a mere technician than her teaching sister, it is also true that every nun is, as such, essentially a teacher. Her vocation gives her the right and the duty to make Christ known to those about her.

If the field of free-time occupations of young grown-ups is more accessible to teaching nuns, that of their attitude to life in general is open to nursing nuns. Their studies and qualifications fit them to organize and inspire Catholic Action groups, or family groups in which questions of family and marital morals can be dealt with: we are thinking of all the evils lurking behind the words abortion, divorce and birth control.

These tasks follow on perfectly naturally from their jobs in the hospitals. It is not so long since nuns were not allowed in maternity wards: their presence seemed inadvisable if not actually 'unsuitable'. Permission to change this was not achieved without a great struggle, but everyone rejoices in the change. Nuns have not been content to stop at opening maternity hospitals but in many of them they are setting up all sorts of enterprises—which we welcome with joy—aimed at educating young mothers to enable them to handle the delicate matrimonial problems which play so great a part in many homes. This is a perfect example of apostolic 'plus-value' which one would like to see copied in many other fields.

The help which nursing nuns specializing in educating young women for family life can give is of inestimable value. These young women have an urgent need to be familiar with Christian teaching on the moral and psychological problems which they constantly meet in best-sellers, films and in everyday conversation. Overhearing the conversation of some of our young people not long out of school, one often has difficulty in believing that they have had a Christian education, so much is their philosophy of life patterned on that of a world steeped in materialism, amorality and neo-paganism. More than ever before they need someone to discuss with them the vital problems the solutions of which will shape the world of tomorrow, and which can only be unlocked with the key of the Gospel.

In the face of the tidal-wave of immorality which is sweeping over those parts of the world which have remained Christian, we need rescue teams; and these teams must be readily available, properly trained, and ready to work frankly and directly. Light must be thrown on the turmoil, and counter-currents set up by the various organs of Catholic Action if souls are not to be swept away in the dark. Our nursing religious must take an active part in this healing work.

The misfortunes of the whole world are our concern, it has been said. This applies to religious too.

They must look beyond the four walls of their convents. It happens sometimes that an institution, say a home for sick children, is situated in the middle of a completely de-christianized area. A large number of nuns is there to care for the sick. Quite right and proper! But could they not set aside some time for the instruction of the laity to re-christianize their surroundings? A religious house is not a world of its own: it must take its place in the world and do its share of the work to be done, while still not losing sight of its main object.

To all women in religion we should like to repeat the motto of the Seigneurs de Gruuthuse, *Plus est en vous*. They have far greater apostolic riches than they imagine. The Church demands that they dare to have faith in all aspects of their vocation, that they dare to believe too in the resources available in the laity. To do something oneself is good; to get other people to do it and thus multiply the activity is better still. We must be obsessed with the idea of a geometrical progression in our contribution to the apostolate. "I have done nothing," one great apostle of our time used to say, "unless I have trained ten others to do what I do and to do it better." If nuns can grasp this, a great step forward will have been made towards the salvation of the world.

9

INTEGRATION IN THE OVERALL PASTORATE

THE term overall pastorate necessarily implies the integration of various elements into a whole. This whole is in the first place the universal Church, and then the local Church and its various ramifications.

Let us start, then, by situating the nun in the Church as such.

Bride of Christ, Bride of the Church

The nun is a dedicated person; she is called to live as a bride of Christ. This is her fundamental vocation and her inalienable glory. This 'marriage to Christ' which is at the heart of her religious profession, is at the same time a marriage to the Church, for the simple reason that one cannot separate the Head from the Body, Christ from His Mystical Body. "The Church and Christ", said Joan of Arc, "are one." Religious consecration in which the nun is vowed, body and soul, to Christ, dedicates her at the same time to the Church. They are one. "What God hath joined together, let no man put asunder." These words of Our Lord apply here too. For the religious, her vows are not something parallel to baptism: the requirements of baptism are realized through her vows in religion. The religious life is but the full flowering of the baptismal vows, the completion of the gift of herself to Christ and the Church. The religious consecrates herself to Christ as He is now living in the Mystical Body of His Church.

94

By this fact alone, the nun's apostolate must be part of a Church-wide perspective. "The world is my parish," aptly expresses the normal Christian view. The nun, that 'high-tension Christian' (if I may use the term), ought to be able to say, "The world is my convent". She has no right to limit her horizon to the four walls of her school or hospital or clinic: it should extend as far as the interests of the Church. A nun has charge of souls, of all the souls she can reach in any way, directly or indirectly, personally or through others whom she trains.

This universality also derives from the fact that her apostolate will be tied in, as an integral part of a much greater whole, with the apostolate of the bishop on the diocesan plane and with that of the Pope, supreme head of the Church, on the universal plane. Under each of these headings it is the salvation of the world that is entrusted to her. Christ must continue by means of His nuns to love souls, all souls, since it was for them that He died and because redemption is for all.

It was to serve the Church that a religious entered this or that particular congregation. The specific aims of her life will vary according to the congregation she chose, but the general purpose is common to all—dedication of one's life to the salvation of the world. In a true Christian perspective, the whole is greater than its parts: it is not the putting together of a number of dioceses that makes the Church, it is the Church who creates the dioceses. It is the same with congregations of religious: it is the Church they must serve; only the means and methods differ. Any attempt to be fundamentally different would be a denial of the primacy of the whole over its parts.

The Church's Bride in Our Times

Loving Christ does not mean loving an abstraction, but loving Him in His living Church as it is today. This is an important truth and the foe of anachronism; it should make immediately obvious the urgent need to live in tune with the Church and the world of our own times. It is in and through the Church of today that the Holy Ghost operates; it is today that we must hear His voice and share His views. *Hodie si vocem Domini audieritis, nolite obdurare corda vestra.* "If you hear the voice of God today do not harden your hearts", the Liturgy tells us. Today is the day we have to meet God.

One does not achieve sanctity in the same way at every stage in history: the road to perfection is charted through the world that is, in the current of graces at present flowing to the Church. It follows that one's first duty to the spirit of a Foundress, who in her time moved within the Church's grace, is to do as she did and conform to the current needs and graces of the Church. The Foundress herself invites—nay insists—that her daughters express the love of Christ which inspired the foundation in a living, up-to-date, modern manner. Not, of course, for the sake of being modern, but in order to be able to correspond to the graces which God gives His Church today.

Attention must therefore be paid to the supernatural trends in the Church, the spiritual Gulf Streams as it were, which appear from time to time and enrich and vivify the Church.

Having seen their place in the Church as a whole, religious must now fix their position locally in matters apostolic.

THE DIOCESAN CHURCH

Our Lord confided His Church, and hence the apostolate, to the Twelve, with Peter at their head, and to their successors. Bishops were placed by the Holy Ghost to guide the Church. Any apostolic action, whatever its nature, must tie in with and extend the apostolate of the bishop.

It is the bishops in their dioceses and in union with the Sovereign Pontiff who are the Apostles [writes Mgr. Renard]. As successors of the Apostles they are charged, as were the first envoys, the first Vicars of Christ, with the task of taking His place after His return to His Father, of founding and building up the Kingdom of God which is the Church. "As my Father has sent me, so I also send you. . . ." "Who listens to you, listens to me." Bishops are, then, by right and by commission, responsible for the apostolate in their own dioceses: the apostolate for the whole of the Church in the diocese, and therefore its organization, is the responsibility of the bishop beyond all shadow of question. Indisputable as the apostolic duty of every baptized and confirmed Christian is, it is the duty of but one cell in a whole organism. Now the head of the whole organism is Peter, and the head of the diocesan organism is the bishop. For the Church which has been confided to him, the apostolate and its organization in all of his territory depend directly from the bishop.[1]

The Bishop and his Priests

But the bishop chooses himself helpers who engage in his apostolate and assist him in it. One does not become a priest just as such, in a vacuum, in one's own way: one is a priest of a certain diocese, attached to the Apostle of the place and in vital touch with him. This association in the priesthood and apostolate of the bishop is more important than individual priestly functions as such. A priest belongs to the Church before he is a parish priest, or a professor or a chaplain; and he is only one of these in order to be able to fulfil some part of the varied episcopal apostolate. Just as a priest shares in the priesthood of his bishop, so

[1] *L'Evêque et son Eglise*, L'action catholique et l'évêque, pp. 152–3 Cahiers de la Pierre qui vire.

H

also he shares in his apostolate. One might apply St. John's words, "It is the shepherd, who tends the sheep, that comes in by the door", in the sense that the bishop is the door, the only permissible entry to any authentic apostolate within the Church of God. Any other entry constitutes house-breaking.

It is in the essence and nature of priesthood that priests are constituted into a college; the priestly body, if I may use the expression, is by its nature dependent from the bishop, stemming from him, participating as a body in his plenary priesthood, making with him and in him only one 'priest'.[1]

Priests and the Laity

If the bishop as head of the local Church must recruit co-workers for himself, the latter must also do likewise and extend themselves by recruiting others: this is the essential relationship between priest and people. The link has dogmatic as well as pragmatic value. The overall work of the pastorate is not in the first place demanded by public order but by the exigencies of the Faith.

Christ is only complete in His members: the Christian is not fully Christian unless he be bound together with other Christians. If this is true for everyone who has been baptized, how much more true must it be for those souls who are dedicated to God.

The idea of the priest as a complete entity in himself has been developed too far: too often has it been said that having recourse to lay assistance was due to shortage of priests and not, as it should be, part of the very nature of things.

The idea of a priesthood cut off and isolated from the

[1] B. Bazatole, *L'Episcopat et L'Eglise Universelle*, L'Eglise au sein de l'Eglise locale, p. 344.

laity is theological nonsense even before it is a handicap to any effective apostolate. A priest must be attached to the laity or his apostolate is paralysed. His ministry loses, even before it starts, all missionary character, for one obviously cannot be a missionary in a vacuum.

A priest whose function is not extended by the co-operation of the laity is an anomaly, a contradiction.

Lack of co-operation not only brings a real danger of the priest being isolated and tempted to despair in the face of the size of his task, it also threatens to paralyse the laity by preventing them from giving of their best.

We must constantly come back to the underlying theme: Christ wanted to attract members to Him for His work of evangelization. In our apostolate we must respect the logic of this fundamental wish.

It follows from all this that a priest is only fully a priest when he is head of certain members united to him in a union which does not, however, suppress initiative or legitimate autonomy. In this extension of the priestly function the first to enrol are those who are already dedicated to God, so that with Him they can bring this vital concept to the laity.

Priests and Religious Communities

Much has been written about priest–laity relationships, but not enough has been done to integrate into this concept the proper place of the nun, who is the immediate extension of the priesthood without however constituting a barrier between the priest and the laity. She must take her inspiration from our Blessed Lady, to whom God Himself gave her role of mediatrix.

Mary is the intermediary between Christ and us in that she helps us to unite ourselves with Him; she does not interpose herself to isolate us from Him; she wants to unite us, bring us nearer to Him and introduce us into His presence.

Mary is the means chosen by God to ensure that our approach to Him is more certain. Mary's mediation is not something supplementary, parallel to the mediation of Christ, it is in the very heart of His unique mediation. The Church invites her nuns to be responsible for the female laity; they have the task of spiritual motherhood. Leaving to the priest his specific functions at the altar and in the administration of the Sacraments, the nun should contrive to be the spiritual inspiration of lay-women. This is not the task of the parish curates, but of the nuns.

A parish is not well balanced pastorally unless nuns complement the work of the priest. Religious education, just like all education, needs their help. There should be a 'mother' in the parish household, and too often the post remains vacant. The lay-women of the parish should be able to fall back on nuns for support in the same way as the men rely on the priests, who are near to them. Particularly in the mission field there are many ways in which nuns can, within limits, replace the priest: for instance by organizing a Sunday service in places without priests.

A Superior General wrote to us on this matter:

> People complain bitterly about the shortage of priests
> . . . and in the mission field it is really terrible. One
> solution seems indicated: make the maximum apostolic
> use of the devotion of nuns.

In the same spirit, H.E. Cardinal Cicognani, Secretary of State to His Holiness, wrote some months ago to the French teaching nuns on the occasion of their Sixth National Congress:

> It is not a matter of indifference to the general good of
> the Church that women called by God to the religious life
> and more particularly dedicated to parish work should

enter fully into the apostolic aims of the parishes where they are helping. Even from the merely human point of view experience shows that a group achieves its full efficaciousness only if there is identity of views and action among its members. This is the case in a parish where the clergy, religious of both sexes, Catholic Actionists and all the faithful work together on the pastoral work, each carrying out the task proper to his or her state.

Just as the priest has too often been pictured as separate from the laity, so it is often with nuns. One hardly thinks of the help that nuns would receive if they meet fully the demands that Christ makes on them. Each one of them should have a group of disciples about her. This, of course, supposes that they have grasped what the role of the laity in the Church is, and that they firmly believe in the royal priesthood of which St. Peter spoke.

It has rightly been said that redressing the pastoral balance must begin by 'converting the priest to the laity'. Congregations of religious need a similar conversion: they must believe in the priesthood of the faithful, in an apostolic laity. Thereafter they will better understand the capital importance of the part they have to play in inspiring the laity.

Nuns and Lay-women

The appearance of the lay apostolate in the organized form in which we know it today is something new in the history of the Church. It is natural, then, that this should start a new chapter in the long history of the apostolic advancement of nuns. Where the apostolate expands, so does the role of the religious. The creation of Catholic Action in all its various forms immediately raises the question of where the place of the nun is in all this.

Here is what Mgr. Renard has to say on this:[1]

Our times have seen the birth and growth of Catholic
Action and the increase in importance of the part played
by the laity in the Church. For the last ten or fifteen
years nuns have been receding into the background,
eclipsed by the fortunate upsurge of a laity courageously
accepting its responsibilities in the City of Man and the
City of God. Even the clergy have sometimes been
affected by this lay activity. Some priests have stiffened
to the defence of their rights and have tried to stem the
rising tide of men and women who, as baptized Christians, wanted to have their say and play their part. . . .
This same feeling of inferiority and envy in relation to a
militant laity has, very humanly, appeared here and there
among nuns; they remember their own ardent, enthu-
siastic and militant youth when they were at grips with
the world, discovering nature and the world in a spirit of
freedom and dedication; they admire and envy the girls
of today and their freedom, their being able to carry on
their work till all hours of the night, taking part in lively
study groups, discussing and revising their approach with
other fighters like themselves, living as part of a team—
while they themselves, the nuns, are tied by their Rule,
dependent on their Superior, hampered in their work by
the habit and the Rule, prevented even (they think) from
achieving their desire for poverty and abandonment to
God by having a comfortable convent and security for
the future. In short, the religious life seems like a cage:
they are protected, of course, as in a fortress, but they
cannot go out and fight the enemy where he is—outside.

These remarks are important. We must guard against this
sort of inferiority complex which threatens to devalue the

[1] *Vie Spirituelle de la religieuse d'aujourd'hui*, pp. 12–13. Desclée
de Brouwer, 1960.

religious vocation. To eliminate it, it is necessary that religious realize the magnificent part they have to play in the renewal of the Christian laity, in this work of 'putting the Church on a missionary footing'.

The emphasis has been put back on the apostolic duty of each baptized Christian. The laity of today is better able to understand Lacordaire's definition of a Christian as 'a man to whom Jesus Christ has confided other men'. But baptized women will not fulfil their task nor awake to full apostolic consciousness unless nuns grasp their role of providing inspiration and stimulus for them. Note that I do not say 'direction'; the direction of the lay apostolate lies in the hands of the laity, but their spiritual inspiration is the responsibility of the priest, and so of those who function as direct extensions of his office, that is to say nuns, his spiritual lieutenants. A nun is not a directress but a teacher, training those responsible for carrying out the tasks. How very desirable it is in any case that the instruction of young women should be in the hands of a nun rather than in those of a young curate, for to reach the necessary stage of instruction requires many prolonged contacts between teacher and pupils. It has not been sufficiently stressed that the absence of religious is responsible for the delay in the general mobilization of lay-women which recent Popes have demanded.

The religious life cannot attain full efficacy unless it be rethought in terms of this new dimension, the laity.

There can be no doubt that every congregation whose object is the active apostolate is already expending its zeal on some task normal to its constitution, whether it be teaching or nursing. Within this duty of their state—this 'professional' duty, one might say—there is plenty of room for great devotion, but this duty also includes as an integral part of the religious vocation the further duty of inspiring the laity. In plain terms this means that it is the nun's job to take part in the triple task which Pius XII set before priests when he said they must discover, train and make use of lay apostles. There are tasks there of recruitment, of instruction, of training, of getting things going, of organizing things which are worth thinking about. Nuns must know how to train the laity for its tasks, how to start movements or put life into them, how to co-ordinate various activities to ensure the highest yield. The dominating thought must be to find people who will 'multiply'.

One must be obsessed with the idea of making the 'apostolic most' of the time available in order to train others.

There is a subtle temptation to rest on one's laurels. We are easily satisfied with the good—the perfectly genuine good—we are doing, and forget about what remains undone, about all that we could and should have turned to apostolic profit. One must always come back to that phrase of Lavelle's, "The greatest good we can do to others is not to display our treasures to them but to show them their own".

A nun does not respond to the greatness of her task unless she is obsessed with the idea of rousing up and training the great mass of lay-women.

In conjunction with the priest she must accept responsibility for the salvation of souls. A convent can never be an island among the waves but must be a promontory jutting out towards the high sea.

Addressing the nuns of his diocese, the Archbishop of Toulouse, Mgr. Garrone, wrote:

I wish you all an ever-increasing desire to co-operate in the general life and apostolate of this diocese where your mission has placed you. You must integrate yourselves fully and without confusion, contributing your personal value and your value as religious. You must be aware of the surroundings in which you find yourselves.

You must be ever more and more united in your common effort, working together fraternally and effectively without overlapping or confusion. Should one congregation among you live its own little life apart from the rest, without reference to the general good or the present needs and activities of the Church, it will have no vocations: and that will be a good thing.

Complementary Truths

When a new idea makes its mark, it often happens that complementary truths take a back place. This is what happened when emphasis was, very fortunately of course, put on the advancement of the lay role in the apostolate. The time is now ripe to round out the picture by including the advancement of the nun's role among the laity.

Some years ago it may have seemed that 'life in religion' and Catholic Action were mutually exclusive. In order to get the lay apostolate going, it was necessary to insist on their own responsibility for it, on the mandate they had received from the hierarchy. The part to be played by religious seemed to be outside all this. All that was asked of nuns was that they should find out what it was all about, give some encouragement from the side-lines, and pray. The idea of active co-operation did not come up even in important official documents. To quote one example, the proceedings of the Fifth Provincial Council of Malines in 1937 merely recorded:

At every opportunity educators will draw the attention of their young charges to the splendour and the necessity of lay participation in the apostolate of the hierarchy, and will exhort them to join now, and later, organizations appropriate to their age and condition. What we have just said in general terms about masters also applies in its entirety to men and women in religion who are engaged in the education of boys and girls. They must take care to inform themselves about all aspects of Catholic Action. Special courses arranged in various places with this end in mind are strongly recommended.

That was how we saw things at that time. A positive part for nuns to play had not been envisaged. We do not intend to analyse the causes of this state of affairs, but only to record a typical fact.

In practice the instruction of militant apostles or even Catholic Actionists was rarely entrusted to nuns, from whom the only collaboration asked was material assistance in as discreet a manner as possible. From there it was but one step to believing that nuns had no mandate for the apostolate.

Papal Directives

Since that time, Pius XII himself gave nuns associated with Catholic Action in Italy a clear statement of what he expected of them, on 3rd January, 1958. He asked that nuns should collaborate with the clergy in the education of young women.

Though it is true that the priest at the altar, from the pulpit or in the confessional, must contribute to the spiritual education and wisdom and prudence of young women, since they are souls committed to his care, it is also true that he must find in you the indispensable collaborators who live in close contact with these girls and can help, support, comfort and console them. The Church is therefore counting on you as direct intermediaries whom the priest can use for the education of young women.[1]

[1] R. Carpentier, S.J., *La Vie Religieuse—Qu'en pense L'Eglise?*, p. 212. Paris, 1959.

He asked them to train apostles:

> Are the pupils in your schools ready, each in the field which Providence will assign to her, to collaborate in the reconstruction of the world? . . . To train young people to look at the world with Christian eyes, to see it as it is, to know what it should be like, and to work towards its conforming to God's plan, these are the practical aims of every Catholic educational or instructional institute.[1]

He asked them to train leaders:

> In this atmosphere of intense training we see the providential birth of the Catholic Action Association. What needs to be done is to place in the souls of pupils a leaven of overflowing life and courageous action; to place at the head of the others a group of resolute pioneers who are capable of swift action and of carrying with them any who may be tempted to lag behind or to fall out. These will be your collaborators in the difficult task of educating your pupils to be Christians. If your local Association can create this leaven, it will not only be good for the health of your Institute but it will also constitute an excellent school for those who will later become leaders.[2]

Much has, quite properly, been made of the mystique of the lay apostolate; just as much has been made of the mystique of marriage, but this should not lead us to ignore that of the priesthood, nor that—still to be created—of the nun as inspiration of the laity, promoter of lay vocations among women and chosen collaborator for the setting up of Christian homes. There is too great a tendency, even in parishes, to relegate nuns to inferior tasks which could as well be done by others. Their place is not just in the sacristy, or looking after children and sick people. The ordinary

[1] Ib., p. 215.   [2] Ib.

healthy grown-up female population has a crying need for their help.

Two decrees of the Roman Synod promulgated by Pope John XXIII make clear the extent to which the Holy See is preoccupied by the integration of nuns into the overall pastoral picture. The first emphasizes the breadth of their role.

> Since all religious strive for the highest and most perfect virtue, let them remember that their vocation is wholly and fully apostolic, is no way bounded by the limits of locality, of things, or of time, but extending everywhere and at all times to anything that touches the honour of their Spouse or the salvation of souls.

The second deals in particular with their part in the organization of lay-women

> As regards women's associations (of Catholic Action) and particularly those of young women, it will be useful to call upon nuns to complete and perfect the task of the chaplain upon whose job prudence places certain limits. Let Superiors assure themselves that their nuns are suitably trained for this form of apostolate; let them freely accord permission to attend congresses and courses with this object; all these means of training being, of course, subject to approval by the ecclesiastical authorities. (Art. 649.)

"Come to Us"

The world of women ought to appear to a nun's eyes as a mission field entrusted to her. The *Acts* describe how St. Paul heard a voice saying, "Come to us in Macedonia", and started out in order not to disappoint the Macedonians. This same invitation comes today from the depths of the unconscious of a spiritually undernourished world, and it is addressed to nuns. We cannot remain deaf to this appeal.

"What others expect of us," says Bernanos, "God expects of us."

One example of a large-scale pastoral operation in which nuns played a notable part has recently come to us from a large Latin-American country. One district had become infamous in the last dozen years or so for the number of assassinations committed there. Two years ago, taking advantage of a period of comparative freedom from violence, a simultaneous mission was organized in all parishes of the diocese. Drawing on every part of the country, teams were set up of a priest, a brother or seminarist, a layman and two nuns. These teams, with a perseverance equalled only by their courage, visited every single household. Two years later, commenting on the success of this unique form of apostolic mission, the bishop said, "These toughs—these *bandoleros*—confessed first of all to the nuns, then they went and asked the priest for absolution".

This example, exceptional though it admittedly is, shows how wrong it would be not to have the courage to believe in the apostolic fruitfulness of nuns. Everyone must be convinced of this, and follow it to its logical conclusion.

The Duty to Increase

If we really want to produce an answer commensurate with the needs of souls, it is of the greatest importance that the clergy co-ordinates its apostolate with those of nuns and lay people, and that everyone thinks all the time in terms of increasing and multiplying.

For a nun to teach the catechism to twenty children is a very good thing, but it is more important that she should train other adults to teach catechism in their turn. One must always harp on the idea not of doing things oneself alone but of making it possible for more people to do them, controlling and supervising them of course, so that apostolic action is further and further extended. I have no right to spend my time producing a 10 per cent. return if I can make it earn 100 per cent. We only have to imagine for a moment what the power and range of the Church's activities would be if every nun in the world were aware of the necessity for her to inspire a group of lay people; if each could imitate our Saviour in sending disciples out in pairs to the missions. I am thinking, for example, of the literally tenfold work of a nun capable of animating a Praesidium of the Legion of Mary. I use this as an example, since the Legion exists in all five continents, because it is very easily handled and its supernatural harvest is exceptional and manifestly blessed by God. But whether it is the Legion of Mary or any other Catholic Action movement, the principle remains and should be accepted by all nuns.

In writing the Life of *Edel Quinn*, that heroic young woman who at the cost of her own health went and founded the Legion of Mary in British Central Africa, I often thought of the paradox presented by the example of a young lay-woman arousing some thousands of apostles in her passing. What a harvest there might have been if each missionary nun had known before embarking for the missions just how to arouse and organize a lay apostolate.

To clarify this idea of multiplication of one's personal contribution, I like to recount a conversation between two Mothers General about helping the clergy. One related how the parish priest of her parish had asked her one Christmas Eve if the nuns could help by singing the Nativity Mass in the parish church. She agreed, but the next day had asked the priest not to make the request another year on account of the inconvenience caused to the community. "That's what you would have done, isn't it?" she ended. The other one, however, had grasped the principle of multiplication and answered, "I should have accepted, as you did. And then, as you did, I should have asked the parish priest not to repeat his request next year. But then I should have

added, 'But, Father, if you would like us to go round the
parish and find you some volunteers to start and train a
choir, we shall willingly do so'."

The answer lies there. And the example is valid in a
thousand variants. It is a matter of getting the right viewpoint—a viewpoint that is much rarer than one imagines.
People think only of what they can do themselves, only
rarely of what they can achieve with others. Yet the salvation of the world depends on seeing things like this. If we
are not to be completely overwhelmed by the magnitude of
the masses needing christianization or re-christianization,
we must be able to apply this basic principle in practice.

The Place of Nuns in the Pastoral Structure

Once nuns are conscious of the scope of their mission
and have by right assumed their place in the general pastoral picture, it is obvious that they have a place at all
levels of the pastoral structure: parish councils, diocesan
unions, national councils and, one day no doubt, international councils.

In the letter we have already quoted, written by Cardinal
Cicognani to nuns in the name of H.H. Pope John XXIII,
express mention is made of the presence of nuns at parish
councils.

The temptation sometimes experienced to treat nuns
as no more than useful assistants for the less important
tasks, must be opposed by the conviction that they are
first-class propagators of the Gospel and very often constitute a necessary link between the pastor and his flock.
This being so, why should their voices not be heard in
councils where pastoral work is planned, as is the case
already in many parishes? Associated thus with the apostolic decisions of the clergy, they cannot fail to put them
into better effect in the exercise of their duties, be they

liturgical, catechetical, educational or medical. The clergy and the parish will gain by it, and in the same way the nuns themselves will be in a better position to develop their vocations among a laity whom they know.

This by no means implies that a congregation of religious should renounce anything of its own specific objects, nor that the community as such has to be integrated into the parish context; it implies, however, that one or another of the members must act as a link between the convent and the parish or diocese and serve as a coupling in the co-ordination of the whole. Without this the community would not be participating in the duty common to all men, in the unity of action which is the concrete, visible, dynamic expression of the mystery of Christ living in His mystical body.

The mere acceptance by nuns of this broader view of which we have been speaking will mean that their action will reach the parents of their pupils, the relations of their patients, their old pupils, the laity—in short, a number of people living in their parish who can be reached in this manner. We are aware that it is often a floating population, difficult to identify with any particular locality; this difficulty is due to the conditions of modern life and serves to demonstrate that to parish apostolic work must be added pastoral work in terms of sociological strata in a much wider field. This sort of apostolate is much easier in the country than in towns and cities. In any case, there is no avoiding the issue; each community must somehow make its contribution to the general work of the pastorate.

In 1949 the cardinals and archbishops of France declared, "It is important that the clergy are able to make nuns enter into the stream of parish life and do not deprive them of their apostolic responsibilities".

The bishops of France, in plenary session in 1960,

speaking of the necessary change-over from a conservative to a missionary pastorate, added, "This demands bold revisions in habits of thought and action: the Spirit of God is there to inspire everyone—priests, religious and laity—with the necessary clarity of vision and courage".

Nuns, then, have their place not only at parish level but also at diocesan or national level, everywhere in fact where women are engaged in the lay apostolate. They have an active part to play in contributing to meetings and congresses the particular treasure which is theirs.

One cannot but greet with joy the notion of such an integration of religious into the work, for it will be of the greatest advantage to the entire Christian community. This mutual aid in the apostolic field brings to its full life and reality the union of all Christians who share in brotherly love the same Eucharistic Supper. To communicate together at the same altar leads naturally to communion together in the same apostolic work—a work which is a unity, yet diversified by the gifts of each worker. What a magnificent opportunity of offering the world the shining example of full Christian brotherly love, for surely the *cor unum et anima una* are still the sign whereby the world shall recognize the disciples of Jesus.

It is clear that in order to be able to respond to these needs, nuns are going to have to adapt themselves, if only because their work with the laity will have to be mostly in the evenings or at week-ends. More important still will be the necessary consideration to be given to a training fitted to the scope of the work to be done. The following chapters are devoted to a consideration of the necessary measures.

I

10

NECESSARY CHANGES

THE nun's job in the world of today cannot be accomplished without certain modifications to the customs of the classical type of religious life.

It is not surprising that as the Church became aware of the growth of the field of its apostolate she issued many appeals through her Popes and bishops for modification of the religious life to suit the new needs.

Such modification, needless to say, does not touch the essence of the religious life nor the unalterable kernel inherent in its dedication to God. It concerns only bringing it into line with the current needs of the apostolate. And it goes without saying, too, that these modifications are not the responsibility of individual nuns, but must be made by competent authority if we are not to have disorder and anarchy, or warp the religious spirit, or just scratch the surface of the problem.

THE POPE'S APPEALS

With these reservations in mind, one should achieve the proper dispositions by re-reading the urgent exhortations of the Popes, particularly Pius XII. It is worth recalling two passages from an address by Pius XII to religious teachers on 13th September, 1951.

It is possible that certain elements of the time-table, certain regulations which are not just interpretations of the Rule, certain customs perhaps more in keeping with another age but which nowadays only complicate the task of education, should be adapted to the new circumstances. . . . Your aim is to serve the cause of Jesus Christ and His Church according to the needs of the world of today. It would therefore not be reasonable to persist in customs or ways of doing things which hinder this service or even perhaps make it impossible.

This invitation to do some pruning is completed by another exhortation to be of their time.

With the help of the Holy Ghost, spirits must be revived and renewed so as to be able as far as possible to cope with the way of life of our epoch and with its spiritual distress.

(12th November, 1950)

Echoing the repeated appeals of the Holy See, many bishops all over the world asked that the necessary changes be made to meet the apostolic needs of the age.

These appeals to our various congregations to place themselves more widely and more squarely on a missionary footing came up against a number of difficulties. But one must repeat with Newman that a thousand difficulties do not constitute a doubt. Obstacles are not insurmountable. All that is needed to get over them is in the first place the spirit of faith to reveal clearly the imperative necessity of the end in view, courageous perseverance, and finally a sense of humour to enable one to keep one's sense of proportion and keep smiling in spite of the difficulties. If it is possible to say *Sacramenta propter homines*—Sacraments exist for men and not the other way about—one should also be able to say, with even greater force, *Regulae propter salutem mundi*—rules are made for the salvation of men.

In this chapter we shall run through the main points

which may be thought to constitute handicaps to missionary zeal.

The necessary changes affect a number of aspects of religious life, among them :

the balance between prayer and the apostolate;
the allocation of time;
the balance between the exigencies of enclosure and those of the apostolate.

I. THE BALANCE BETWEEN PRAYER AND THE APOSTOLATE

Praying by Action

A proper balance between prayer and apostolic action is essential to a life vowed to the apostolate. This balance is easily obtained if prayer is understood in depth and the apostolic activities are truly apostolic. Prayer is not an island in time unconnected with the rest of life, and its value is not measured by the clock nor by the multiplicity of exercises. Our Lord did not say we must pray for an hour, or two hours, or three : He said we must pray always. This supposes a life of prayer, a permanent state of communion with God. Seen in that light, as a state of union, prayer is uninterrupted : this is the essence of the matter. The true apostolate is prayer in action. Note that it is not the action as such that constitutes prayer, but supernatural apostolic action. One is right to be on one's guard against the 'action heresy' : that refers to purely human action, merely being active, on the level of purely human techniques. The supernatural action which is the apostolate we are talking about consists in communicating life and grace, in giving Christ to others. This giving is spiritually vivifying in itself and also in its returns. There is no discontinuity between prayer and the worship of God on the one hand, and on the other, the work of evangelization in the strict sense where one shares the theological virtues of faith, hope and charity with others. The apostolate is a supernatural work of extension of the Redemption. Quite naturally it will be steeped in prayer before, during and after the activities which it requires. Discontinuity only exists between superficial prayer and action which is purely human.

In the logic of our Christian life we must not only turn our prayers into action but also turn our actions into prayer.

Turning our prayers into action means embodying in the apostolate, and by means thereof, all that we ask for in prayer; it is what transforms prayer into action and is the essential complement to the *Thy Kingdom come* of the Our Father. The apostolate is the outcome of true prayer, and its guarantee and touchstone.

And inversely, we must 'pray' our actions right from the start, for 'without God we can do nothing', without Him we are totally incapable of achieving good.

Spiritual Exercises

In practice a very large part of the life of prayer is devoted to 'spiritual exercises'. These exercises have a very important place in the organization of the religious life.

One must start by putting in a class on its own the supreme meeting which daily unites God and man: the holy sacrifice of the Mass and the Communion which is its culmination. This is obviously the peak of the spiritual life of religious. All that has been done to bring before the laity the value of the Mass is eminently applicable to nuns. Their piety must nourish itself at this primal source. The liturgical revival is as much a boon to religious houses as it has been for the laity: they cannot remain outside it, for their own sakes as well as for that of their pupils. They must profit, and see that others do too, from the instructions issued by the bishops about the introduction of more solemn, dialogue Masses in their communities; they must see that their pupils take an active part in them so as to prepare them to take their place in their parish communities.

Spiritual exercises serve as a frame for the Mass thus lived and brought to life and prolong its action through the day. These exercises cannot be left to individual initiative, but they should be restricted to what is essential, and should really constitute communal prayer. There is a need to amend and simplify them, to give their piety a more biblical, liturgical, ecclesiastical and apostolic bias. At the start of one's spiritual life it is reasonable that much time is devoted to vocal prayer and communal exercises designed to instil the habit of prayer.

Spiritual progress will make more time available for personal prayer. It will be as well to allow for some freedom in these exercises and to reconsider their distribution which tends at present to divide the day into too many short periods. Certain out-of-date and redundant devotions— and there are enough of them in all conscience—must be mercilessly eliminated. Too numerous and ill-adapted spiritual exercises tend to make the life of prayer mechanical and to atrophy it; they can become an end in themselves instead of remaining a means. By contrast, properly understood they lead gradually to a permanent state of prayer and communion with God. This sort of result will come particularly from short periods of interior concentration which will punctuate the day like resting places where one can breathe deeply of the air of the supernatural and unite oneself more consciously with God.

It is difficult to see active religious being bound to sing office in choir. St. Ignatius showed the way when, for sound apostolic reasons, he omitted certain practices more suitable to the contemplative Orders. And would it not be as well to have a look at the musical side of offices sung in a manner irritating to modern ears? It is hard to imagine a modern girl being attracted by the high-pitched quavering of certain tones, and one would like to see a more natural and relaxed mode introduced. The singing of Lauds and Vespers, the liturgical prayers *par excellence*, in choir could well suffice for active nuns and would ensure their participation in the great prayer of the Church: the rest of the office could be voluntary and left to the discretion of the individual.

Spiritual reading for professed nuns could be spread out through the week (not through the day) so that the same number of hours could be devoted to it without cutting both the day and the reading mechanically into sections. No doubt the nuns would thereby derive greater benefit from the reading. The examination of conscience and the chapter of faults should be reviewed in the spirit of the times and perhaps benefit from current practice in certain movements where they have a 'review of life'. One should aim at more candour and openness between nuns, more mutual sincerity, more real sharing, at constructive self-criticism in common, so as to help build up a great work. This means that religious usages as well must move in the missionary dimension, that one should question oneself before God, together and singly, about the answer that has been made to His invitation to each individual to share in the salvation of the world.

All these observations tend to the creation of a more supple and open conception of the religious life, giving rein to the Holy Ghost and to the action of grace in souls, and helping souls to embody in their lives the essential one-ness of prayer and the apostolate.

2. ALLOCATION OF TIME

How, one might ask, are the supplementary or increased tasks put forward in these pages going to be fitted into an already overloaded time-table? Look how much there is to be done in one day, count the numbers of demands for their services . . . and have pity on the nuns!

Scale of Values

This is a major difficulty and must be examined with the eye of faith. There is a theology of time, just as there is a theology of work. Time must be used and filled in accordance with the scale of values which a sincere and generous faith reveals. Which brings us back to the fundamental question: Why were we created? We are created to know God and to make Him known, to love God and to make Him loved, to serve God and make others serve Him. This is the overriding consideration for every baptized person; all the more so, then, for a religious solemnly vowed to God. She must be obsessed by time. We have not a minute to lose when the glory of God and the salvation of the world are at stake. This means we must review, and review again, the use we make of time. *Caritas Christi urget nos.* We have no spare soul, no spare life, and it is to this generation that we have to convey God's message. This does not rule out relaxation—which is essential for good work—but is just a reminder of the precious value of time. Time is money, the saying goes. It is true. One could say with Newman, "Time is eternity", which is the same as saying that time is souls. The way our day is filled must be rigorously examined in relation to this scale of values.

Everything is grace, for sure; everything can contribute to redemption [wrote Fr. Courtois]. But at a time when there is such a shortage of men and women to work in the Father's field, faced as we are with the immense tasks of the apostolate requiring, as the Sovereign Pontiff has himself made clear, the maximum effort from everybody, it is vital that a considerable part of the time at the disposal of nuns should be spent, under obedience of course, in direct evangelization.[1]

This call to the duty of evangelization, echoing St. Paul's "Woe to me if I do not preach the Gospel!" is also in a certain way addressed by the Church to contemplative nuns. We know that the Sacred Congregation for the Affairs of Religious on 19th March, 1952, on the basis of the Constitution *Sponsa Christi*, required of women in religion, including those in enclosed Orders, that they take on certain works of evangelization. "It seems", wrote the Congregation, "that the time has come to reconcile the monastic life, including in a general way that of nuns dedicated to a contemplative life, with an appropriate degree of participation in the apostolate."

Our scale of values demands that God be served first, and that priority be given to serving Him as directly as possible, through one's neighbour. Supernatural things are in a class of their own. One must give one's time to God, and give God to man, preferably personally when this is possible. Given these principles, the time-tables of communities must be arranged accordingly. It is in terms of God that one must start making the changes. The earth revolves round the sun, not the sun round the earth : we must accept Copernicus' teaching—and apply it to God, the sun of our existence as individuals or as communities.

The problem is one of balance, of proportion. Professional duties will continue to take up the greater part of the day : in this, nuns are applying in practice the principles of Christianity and are working (and very effectively) for the salvation of the world. They must make the most of all opportunities for evangelization occurring within the framework of their professional duties. But they must also set aside some periods for the organization, inspiration and

[1] *L'Heure de Jésus*, Ed. Fleurus, p. 133–4.

guidance of lay-women who, having had this training, will in turn be able to introduce the practical requirements of the Gospel among those with whom they work or spend their leisure hours. Without the assistance of nuns, the female laity will never play the essential part desired by the Church; without them it will never be organized on the scale desired nor inspired to undertake the sort of 'border raids' outside Catholic circles which the Church expects of her practising laity.

Nuns must regard the time set aside for this spiritual animation of the laity as sacrosanct. This time, which will have to be squeezed in somewhere in the week, or perhaps only every other week, will necessarily be short—but it must at all costs be found: time-tables must be altered and allocation of duties rearranged until this is managed. This means that community time-tables may have to be altered to suit the times convenient for the laity: the shepherd has to suit his pace to the sheep, not the other way about.

The Threat of Urgency

We must beware of the temptation to believe such adaptations impossible. No one doubts the necessity of eating—and one manages to find the time for it. The same applies to what God asks of us: we must find time to feed the multitudes who hunger and thirst for God and who, directly or indirectly, expect to receive the bread of life from us.

The balance we must find will entail a better distribution of work. Some tasks will be dropped or passed to others The risk that they will be less well done must be accepted. for the balance between spiritual exercises, the apostolate and professional duties must be achieved at any price.

When arranging the allocation of time one must never fail to distinguish between urgency and importance. Unless one is constantly on one's guard, one is perpetually tempted to confuse them, and one can end by spending one's time rushing from one urgency to another, in a permanent state of tension and always having to deal with some new emergency.

One must know how to stand back a bit and give priority where the proper scale of values places it. It is not enough for us to lead 'full' lives regardless of what they are full of : their first charge must be the work Christ demands of us, in the order of priority of the Faith. The temptation to deal with the most urgent things first is a subtle one; to do so enables one to see immediate results and enjoy a quiet conscience and a sense of achievement in spite of the major omissions one is unknowingly responsible for.

As the Very Rev. Fr. De Gryze, Superior General of the Scheut Fathers, wrote to his missionaries :

Hard work can also be a danger : one can do a great deal of work and not give enough thought to one's method of work. Not planning one's work and not co-ordinating it can mean a lot of wasted effort or only half-effectual work. A passion for activity can make one a slave to immediacy and detail, leading to failure to take the large view and to not looking into the future. Too much realism can be fatal, or at least very weakening. We may suffer from lack of imagination, inability to see things in perspective, and loss of creative and recreative thought. We may forget that preaching the Gospel is a process of growth and therefore demands constant returns to source and continual adaptation.

3. THE WORLD AND CLOISTER

Double Demand

The chief obstacle to the apostolic advancement of nuns seems to be the conception of enclosure and what it entails.

There is an apparent contradiction. How can one apply both parts of Our Lord's words to His disciples when He said they were not of the world yet must not flee the world.

To be separated from the world and yet to be present in it to inspire it and save it, is a situation paradoxical enough for all Christians but particularly so for nuns. The apparent contradiction is resolved when one realizes that for active congregations the separation is an attitude of mind rather than a matter of walls and grilles. In her heart the nun has renounced the privilege of founding a home, has left her family and chosen renunciation as her path. But this break with the world is not made in order to isolate herself from the world but to give expression to her greater zeal in its regard. It is the price of a spiritual motherhood opening its arms to all the world's distress. The nun has stepped aside so as to be more effectively present, as the yeast starts by working against the dough in order to knead it and eventually raise it. To isolate the yeast from the dough for fear of contagion is to miss the whole point. By its very nature the apostolate implies a risk, just as the yeast runs the risk of being overwhelmed by the dough. But it is a good Christian risk which makes Christianity an adventure worth living. Without risk even human activities come up against a blank wall. If you exclude all risks of contagion, you exclude care of the sick; if you insist on avoiding all risks from wind and wave, you prohibit travel by air and sea. The Christian apostolate would just be two meaningless words if all risks were barred.

That in no way means that one does not have to take the necessary precautions. Every time an aircraft lands, everything is inspected and checked for the next flight. Incidentally, to maintain its serviceability an aircraft must fly regularly: a regular flight routine is a guarantee that the mechanism remains in a good state.

In the same sort of way, nuns are asked to go to the world to save it.

So much emphasis has been put on this notion of danger that it is important to redress the balance by emphasizing the 'necessary presence' idea. The religious life has quite properly been hedged about with safeguards, sometimes down to the smallest detail. Nobody disputes that the world is full of pitfalls, is pagan and getting more so. One must advance circumspectly, not rush in bull-headed. But when a vehicle is fitted with powerful brakes, one does not hesitate to take it on the road for fear of spoiling the brakes; brakes are fitted to provide safety at high speeds, not to keep one down to a snail's pace. By all means let us make sure that our brakes are in good trim, but then let us take a sight on the horizon and start moving, however hard the road ahead may be.

This dual attitude towards the world—loving it and yet fleeing it—comes from the fact that Our Lord uses the word in two senses. At one moment the world represents the assembly of forces arrayed against God, something to be resisted and fought; at another it is the assembly of men who have to be saved; men in their insignificance but also in their grandeur, in their zeal for good, in the unease of their unconscious search for God.

The argument *Prudence* v. *Audacity* might as well be between deaf people so long as prudence and audacity are assumed to be incompatible opposites. Agreement can only be reached when it is understood that Prudence is the hand-maid of Audacity and that the precautions which Prudence takes are what makes Audacity's advance possible in safety.

Realism in the Apostolate

The separation of religious from the world cannot be such that a nun can no longer play the part of mediatrix between God and the world. One must avoid giving the impression that nuns live in a sort of ghetto, or a hot-house, and that their knowledge of the world is limited to what they pick up from their pupils or patients.

Lay people complain of a certain childishness, naïveté, an old-fashioned air, a too narrow outlook among nuns so that they never see the whole picture and can never really know the world. Cut off from the world, the nun is in danger of not getting a realistic picture of the spiritual ills that afflict mankind. One must enter on the apostolate with open eyes : being an ostrich is the worst possible approach.

Without going into details about rules of enclosure, which everyone wants to see simplified in the revision of the Code of Canon Law which will follow the Council, it seems appropriate to indicate the basic principle which will probably guide its provisions. This principle is that woman can no longer be treated as a minor. The practical corollaries of the principle must also be accepted. We do not in any way deny the differences between the sexes, but in the service of God and man there is a substantial degree of equality in all dedicated souls. In principle, therefore—and leaving aside the priesthood and certain special cases—women in religion should have the same treatment as men in religion, due regard being paid to feminine psychology. Women are no longer minors and one no longer has the right to consider nuns as unable to undertake enterprises that their sisters in the world undertake in many different spheres of modern life.

Enclosure and the Apostolate

Everything we said in Chapter 8, 'The Wider View', would be nullified by a rigid conception of enclosure. How would it be possible to meet all the needs, to make the necessary contacts with families in their own homes, to ensure the spiritual stimulation of the laity, to achieve the realism essential for any 'made-to-measure' apostolate, if teaching and nursing nuns remained confined *intra muros*? It would be equivalent to entering a plea in bar against the Church's claim to missionary zeal; it would be to mistake entirely the scope of the duties of their state for the active congregations of today. One example will suffice to show how a rigid conception of enclosure threatens today to paralyse teaching nuns in the carrying out of their primary mission.

When a congregation, whatever its original nature may have been, agrees to take on the job of educating young Christians, it binds itself *ipso facto*, in justice to the children, to the parents, to society and to the Church, to do the job fully and properly and to adopt all necessary means to do so. Now, education today requires the presence of nuns in various departments which did not exist yesterday. But the congregation undertook total education. This means that the mistress cannot be separated from her pupils by some enclosure rule which forbids her presence where it is necessary to fulfil the mission entrusted her by the parents and the Church. We know the influence that their spare-time activities have on children, how these out-of-school activities can mark them for life, and the risk there is of undoing in the course of one week-end or one holiday all that has been achieved at school. This emphasizes the necessity for extra-mural contacts with young adolescents who need training in the various apostolic movements, guidance for their study-trips, and social, family and cultural formation.

Besides this out-of-school world, there is the day-to-day reality of life which demands the presence of the nuns with the children or with their parents in the case of bereavement, or accident, ill-health or suffering, to give only a few examples. To hand the pupils over to the lay staff in a series of such circumstances because the rules of enclosure forbid nuns to leave the convent, is to cut off the human and Christian contact at the very moment when nuns could best fill the important role expected of them.

And unless enclosure rules are made more flexible, how can nuns fulfil their mission to the laity alongside whom they live—parents, lay staff, old girls, etc. They can only do so if they are able to approach those who do not come of their own accord to them. If they have no contacts, how can they manage to train even the best candidates for the apostolate in their respective circles? Apostolic vocations have to be sought out and fostered: they do not normally appear in a flash of lightning, but as the result of repeated contacts and persuasive insistence. They have to be gained one by one. A fervent religious house is like an electric power station, capable of the greatest distribution of power—but someone has to go out and wire up the circuits connecting the outside with the source of power.

There is nothing insoluble about the problem, only one must have the will to solve it. The Decree of 31st May, 1957, allows Superiors much latitude in permitting visits outside which would serve the apostolic work of the house. The only trips expressly excluded are those for personal interest or satisfaction—which is just wisdom. There can never be question of visits to people's homes being transformed into a search for personal satisfaction: the intention must always be to do a service. Instead of making use of the latitude permitted in the matter of trips outside the convent just to acquire new diplomas or arrange matters of finance or health, all one needs to do is to add to the list of accepted grounds that of apostolic motive.

One Mother General, who shares these views, wrote to us:

> Our Constitution says, "No one will leave the convent except on important business of the convent." Up to now the interpretation given to these words has been to

apply them only to trips made necessary for professional reasons or for the material maintenance of our houses. Having now understood that 'important business of the convent' should include anything that helps to achieve our aim of 'making God known and loved', we have not hesitated to include in the phrase all visits demanded by the apostolic needs of our time.

Enclosure and Family

Another question which must be re-thought is that of periodic visits by religious to their families. The religious has chosen a new family; her house is her convent and her fellow-nuns her sisters. Being at home means in the first instance being in her convent, and that is how she sees it. Visits to families must not be regarded in the same light as a boarder's holidays, but rather as a reconciliation of two duties: that of maintaining a certain distance from the world, as properly desired by every religious, and that of *pietas*, family charity—and sometimes even justice—towards one's own people, without mentioning the matter of apostolic presence. For always and everywhere, among her family as much as anywhere else, the religious must be the bringer of Christ; her visible interior joy will be a striking apostolic witness. Her presence in the family home should be a grace for everyone, including herself, for contact with the realities of life will give her more understanding of others and a better appreciation of her own vocation.

Our modern world is more sensitive to the niceties of expression of filial piety than previously. No doubt the younger generation of nuns will appreciate this more than the older ones who might in some cases be confused by what might seem to be laxity. But these latter need only be shown the extent to which these family visits tie in with the general policy of the Church today, and they will

K

accept them whole-heartedly. It will not, then, be as laxity, or worldly accommodation, that they will see these visits, but first and foremost as a filial duty to be fulfilled.

When a parish has a Vocation Day, if nuns born in the parish can come back for the occasion, their presence will no doubt prove a great stimulus for further vocations.

The arrangements for, and frequency of, family visits are things which have to be settled. The customs in force for male religious may serve as a useful guide, but the important thing is to find a balanced and sensible solution, free of petty provisions, which takes local conditions into account and which is open to suitable interpretation by the appropriate ecclesiastical authority if necessary. It will not be possible to modify the rules to cover every case, since cases vary enormously : but Superiors must resist the temptation to stick to a fixed rate of 'one day a year' and must be prepared to accept their responsibilities in full. They should remember what Aristotle said—that true equality consists in treating unequal things unequally—and train their communities to understand the good sense of it.

The Religious Habit

Another, somewhat more obvious, problem arises when there is question of nuns mingling with the world—that of the religious habit.

On account of their aims secular institutes generally do not wear a distinctive habit, but in our view it remains necessary for religious congregations. We have already shown the necessity for a visible and understandable collective witness : some sort of habit which is a distinctive sign is therefore necessary. It has its own nobility, it reminds us of the nun's exclusive dedication to God, makes it easier to keep conversation on the spiritual plane, facilitates confidences, and is a joy and a stimulus for the wearer.

But the habit must be in keeping with the needs of our

time. The world today has no patience with mere ornamentation, useless complications, gofferings and other oddities, whether starched or floating in the wind, which belong to another age: anything contrived or lacking in simplicity is rejected, and anything unpractical or unhygienic, anything that gives the impression that the nun is not only apart from the world but also a complete stranger to its evolution. The habit must be fully adapted to the nun's apostolate: as it is today, it often inhibits her social contacts. In de-christianized circles it acts as counter-propaganda, giving the impression that Christianity is out of date, archaic.

It is astonishing to note the timidity, the inertia, with which those responsible have answered—or rather, have failed to answer—the repeated appeals of Pius XII and other qualified authorities; appeals, incidentally, which are well known and commented upon among the laity. What modifications have been made have generally been minimal; what is wanted is a radical modernization to twentieth-century standards, not to those of some past age. A serious factor in this anachronism is that the visible exterior leads one to fear the existence of an interior inability to adapt to current needs. Some habits exist which prevent nuns from taking part, as they should, in certain educational and apostolic activities, or which embarrass luke-warm lay people and which in many cases are an obstacle to the discretion sometimes needed for house visits.

Some lay people may have a sentimental, historical or traditional attachment to certain habits, but this should be no excuse for giving way to the temptation to keep things unchanged. We must love the Church of today and progress in step with her, according to her wishes and recommendations. This is true even if the origin of the habit lies in some dream or apparition, for it is to Christ living and speaking in the Church today that we owe our loyalty. Saints, and Foundresses of congregations, were models of faithfulness to the Church. One knows how even such favoured souls as St. Teresa of Avila and St. Margaret Mary inculcated in those around them that it was the Church alone, speaking through her human hierarchy, who had in the last resort the authority to interpret the private revelations with which God had favoured them. Loyalty to the Church on the part of nuns is but respect for the loyalty displayed by their Foundresses.

A Direct Manner

Those, then, are the main aspects of revision which interest us to start with. No doubt there are others, less important from the apostolic point of view but which would help from a psychological standpoint to create a better harmony between the religious life and the legitimate aspirations of the contemporary world. There must be, wherever possible, dialogue on the same wavelength : there are secondary aspects of the religious life which would gain by being modernized. Why should there not be a periodical spring-cleaning to get rid of encumbrances and things which are an unnecessary stumbling-block to the modern mentality. The young woman of today likes a clear, direct, simple and frank manner; she should not be put off by an archaic vocabulary or a sort of ritual which is not of our times. These things, small in themselves and of only relative value, have a symbolic importance for our contemporaries : it should never happen that some out-of-date custom should stop a hesitant vocation at the threshold. One cannot hesitate before such minor sacrifices, for it is a question of vocations, which are worth the elimination of a few incidentals.

Let us read again the words of Pius XII which our pages have sought only to comment on and paraphrase :

The art of education is in many ways the art of adapting oneself to the age, temperament, character, capacity, needs and reasonable aspirations of the youth of today . . . adapting oneself to the rhythm of the general progress of humanity. (May, 1951.)

In September, 1952, the Pope returned to the charge in connection with the vocations crisis. He called upon Superiors General to see that "customs, the kind of life or asceticism of religious families did not constitute a barrier or a source of failure". He was speaking, he continued, of "certain customs which, though they formerly had some meaning in a different cultural context, no longer have any and in which a young, fervent and courageous girl would find nothing but fetters inhibiting her vocation and her apostolate".

Adaptation is not Relaxation

To end this chapter on the changes demanded by the needs of the apostolate, one last clarification seems useful.

There must be no mistake: the adaptations we have spoken of do not imply any relaxation or any compromise with the spirit of the world. Nuns who would like to modernize their life and their institutions on those lines have entirely misunderstood our thinking. To adapt apostolically is not to introduce luxury or excessive comfort, nor to follow every craze in order to be right up to the minute, and thus exude an atmosphere of worldliness and superficiality. No, it means quite a different sort of adaptation and one that has as its fruit: a religious life more religious because more apostolic, a more intense life of prayer, a more exacting spirit of renunciation, and a more authentic supernatural spirit. The liberty of action called for is always under discipline and obedience and will be exercised under the control of Superiors aware of the current needs. Far from being an excuse for following one's natural inclinations, apostolic work will be a harsh school of mortification, penance and renunciation. When one knows the self-denial required to start the smallest Catholic Action group, when one knows from experience what it costs to bring just one soul back to God, one knows that 'nature' does not get one very far and one must plunge into the supernatural in order to persevere.

We hope these remarks will prevent any misunderstanding, and will help readers to understand the deeper meaning of these pages.

11

THE REQUIREMENTS FOR
ADEQUATE TRAINING

THROUGHOUT these pages the reader may have been tempted to compare the ideal put forward with certain known realities, and our proposals will perhaps have appeared a trifle utopian. We are perfectly well aware that the incarnation of an ideal is achieved only by a transition which takes time, and that one must take into account the human 'material' and individual and collective psychology.

That is why we must now tackle the problem which is the key to final success: the problem of providing for our nuns a training adequate to fit them to face the current demands of their vocation.

This training must take place in several fields, all of them necessary and mutually complementary. It must be spiritual, apostolic, professional and social. Let us study these in turn.

SPIRITUAL TRAINING

The Novitiate

Spiritual education is given within the framework of the canonical year to which the Church quite rightly attaches great importance and which must be respected to the last letter. It is not difficult to understand; the girl who has just left the world to consecrate herself to God must start by getting away from the world she is leaving. She must discover what she has so far only glimpsed—intimacy with

God. Calling her, the Master said as He did to Andrew and John, "Come and see!" and He now invites her to widen the horizons of her spiritual life. She must learn to know herself, to correct her faults, to discover the meaning and value of her vows and of life in community. The novitiate ought to establish peace in her soul, strengthen her vocation and be a sort of prolonged retreat like Jesus' forty days in the desert before He started His public life. The Church wants her future religious to be Mary before being Martha so that later she can the better be Mary and Martha together in one vocation.

A new world opens before the novice: she has to discover the meaning of prayer, continuous communication with God, life in the presence of God. She must learn to see with the eyes of faith, to hope in the invisible realities, and to love not with her own heart but with that of God. She must enter into the mystery of Christ, grasp the vital significance of the doctrine of the Mystical Body and the Redemption. She must understand Mary's spiritual motherhood which she will have to exercise. She must become familiar with the lives of the Saints, that 'Gospel in pictures'. All this cannot be done in a day, and takes time to mature. The mere listing of these, which are only some of the primary requirements, shows clearly enough that it is not easy to meet all the demands, partly due to shortness of time but also due in part to lack of personnel to do the instruction, particularly in small congregations. Small congregations would profit by setting up, with the necessary guarantees, a common novitiate. But that would not be the whole answer; everyone is aware of the need for greater consistency in the present training and for some supplementary training. In the nature of things the indispensable initial training cannot provide a complete initiation in one year—nor even in two.

Hence the birth of the idea of the juniorate.

The Juniorate

The Holy See is in favour of this idea of the *Grand Juvénat* which would prolong the training period in a different context, and it would like to see the progressive introduction of some years of juniorate.

Fr. Gambari of the Sacred Congregation for the Affairs of Religious explained the object as follows.

> The object of the juniorate is to continue, consolidate and perfect religious instruction both general and particular and also to give the professional instruction necessary for a suitable apostolic activity. The whole should be informed and guided by personal religious growth which is the individual response to religious and professional training.

On the duration, he adds:

> A minimum of two years and a maximum of five should be devoted to the juniorate. Between these limits the period would be settled by each religious family according to its own needs and capabilities.[1]

Nobody can fail to see that this innovation, which is still in its earliest stages, bids fair to make a solid contribution towards making the most out of the religious life.

Some Aspects of Spiritual Education

During the time of the novitiate and the juniorate, the spiritual life of each individual must be very specially deepened, but so also must the specifically religious life be deepened in terms of the spirit of the congregation. The practice of the vows will be decisive in the apostolic education. If the vows appear as means of detachment in order

[1] *Le Grand Juvénat selon l'esprit et les directives du Saint-Siège.* Article in *Supplément de la Vie Spirituelle*, No. 54, 1960.

to make one more available, their practice will, far from being a hindrance, enhance apostolic zeal. Initiation into the practice of obedience will have a considerable influence on the orientation of a religious life. That is why we shall spend some time on it.

Education to Real Obedience

The three vows of religion are very closely related and this relationship is embodied in obedience, the pivot of the religious life.

> What is the religious state if it is not the subjection of the individual, and therefore of his whole life, to the service of God? . . . When he abandons his will to his Superior, the religious gives up to him, and through him to God, all acts that would arise from this will . . . in other words, the matter of the vows of chastity and poverty. Thus one can truthfully say that the vow of obedience achieves the object of the other two vows and so embraces the whole of the religious life.[1]

The central place of obedience in the religious life demands particularly careful training in it. Obedience must be lived as a positive virtue, a stimulus rather than an invitation to passiveness. If it is not properly understood, it will tend to inhibit missionary zeal; properly understood, it will guarantee on the contrary its full freedom. Obedience does not mean lack of personality, initiative or responsibility, but giving to God one's deepest will in order the better to serve Him in the field which authority, His interpreter, indicates. Obedience cannot be productive unless there is openness, reciprocity and dialogue. Submission must not be confused with passivity. Real submission carries out loyally the orders given after having, if necessary,

[1] Dom O. Lottin, *Etudes de Morale, Histoire et Doctrine*, p. 255. Duculot, Gembloux, 1961.

raised the points which seem to favour another course. The most passive nun is not the most subject nor the most obedient.

Although it comprises it, obedience is not in the first instance an exercise in mortification. The prime consideration is not the abdication of one's own will nor the submission to a person, it is loyalty to the common good as an expression of God's will. Everyone must do the job of his station, the Superior as well as the subordinate. And one must never forget that both Superiors and inferiors must be trained in this responsibility which is essential to the success of their work in common. The Superior has the final decision, the last word—but not the second-last word. It is normal and sensible that subordinates be asked for and give their views before a decision is made. The common good demands that this be done with respect and loyalty, and in some cases privately; but thereafter the common good demands loyal adherence to the decisions made.

The proper motive is essential if obedience is not to defeat its purpose entirely. To think that one obeys in the first instance in order to renounce one's own will, is to start on a path that leads nowhere; the primary reason for obedience is not to mortify oneself but to serve God and one's neighbour. One does not obey in order to please someone, but to respond to God's will. Passive conformism is not obedience: obedience has nothing in common with 'complacent paralysis'.

Passivism—passivity as a policy—favours 'no change' and infantilism. As an adult one has not the right not to obey in an adult manner. This does not mean that one obeys any less well: on the contrary. Real obedience is required, based on disregard of self and informed by the theological virtue of faith which recognizes Christ in properly constituted authority. Education in this sort of obedience is something requiring skill and tact which should be carried out by both sides under the constant guidance of the Holy Ghost. It is probably more difficult for a woman to carry it out than for a man. Pius XII himself once remarked to the Mothers General, "What psychology supposes is doubtless true, that a woman in authority does not manage as easily as a man to strike the exact balance between severity and kindness". (15th September, 1952.) To exercise command without stifling the personality of one's subordinates but making on the contrary the most of their initiative and intelligence is the mark of true authority.

Respect for Natural Qualities

Complete religious education does not end with initiation into the supernatural virtues and the vows—it ought also to build up the natural feelings of rectitude, loyalty, justice and social sense.

We must develop in the nun of tomorrow the sense of respect, tact, discretion; she must know the cost of time and the value of silence. She must learn to talk and express herself clearly. She must know how to open out, to share her treasure, and to listen, to be 'pure attention to the existence of another'.

She must develop, suitably balanced, the gifts of the heart; lack of feeling is not a good quality. As Fr. Voillaume wrote :

The love of perfect charity must blossom into a sensitivity which is not only purified but also enlarged and refined, and will permit it to express itself towards God and man with all the treasures of tenderness, friendship, sweetness and strength of a human heart. The perfection of man, who has become the son of God, cannot be dependent on the mutilation of what is most beautiful in him. The heart must therefore not be confined or brutally repressed; it must be directed; its sensitivity must not be destroyed, but purified and brought under control. This is not the same thing, for it is not a matter of lessening and stifling, but of making the heart greater and the sensibility more delicate so that both may be placed at the service of supernaturalized love.[1]

The natural virtues must be developed to provide grace with a fertile soil and because the lay world is particularly susceptible to contacts. Anything that dehumanizes or defeminizes the nun lowers her apostolic value. One must beware of any trace of Jansenism, a one-sided insistence on the supernatural virtues must not be the cause of neglecting the natural virtues.

APOSTOLIC EDUCATION

Spiritual education would be dangerously incomplete if it did not go hand in hand with a progressive theoretical and practical education for the apostolate. The latter must take into account the new conditions of our time. The greatest single new factor is that the laity has become conscious of its apostolic mission in the Church. From which it follows that the apostolic education of nuns would be incomplete if it did not include instruction in the role they have to play as inspiration for the lay-women alongside whom they will constantly find themselves.

The object of apostolic education will be to teach the religious how to give Christ to the world, how to put the Gospel into practice in all its aspects, and how to train others to be in their turn agents of christianization.

The religious has to have a double apostolic training. On the one hand she must be prepared for her task as teacher or nurse with its direct apostolic extensions, and on the other she must learn how to train the laity and thus multiply apostolic action in the world.

[1] *Au coeur des masses*, pp. 376-7.

The Holy See, in speaking of the means of developing the apostolic sense, gives the following directive in Art. 47 of the General Statutes of *Sedes Sapientiae*.

During the entire period of formation and probation Superiors and Masters will not fail to impart a taste for the apostolate to their pupils; more, they will make it their duty to give them moderate exercise in it in accordance with the mind of the Church and the nature and objects of the institute.

By the same token, let us not forget the wish expressed by the World Congress of Major Superiors in Rome as reported in the *Osservatore Romano* of 16th December, 1957.

With regard to non-priest members, whether male or female, of religious institutes, Congress expresses the wish that special attention be paid to the indispensable part these persons have to play in the creation, training, organization and apostolic inspiration of the adult laity. It is their task, under the direction of and in close harmony with the clergy, to collaborate in the work of setting the entire Church on a missionary footing, associating themselves in the threefold task commended to priests by Pius XII of discovering lay collaborators, training them, and making use of them in order to increase the apostolic yield. [Letter to Lenten Preachers, 1954.] This role may be compared with that of N.C.O.s in an army, where they are an indispensable element of liaison and co-ordination. So that they may more effectively fulfil this very important aspect of their religious vocation, a practical introduction to the apostolate seems highly desirable with a view to the instruction and training of the laity.

This progressive introduction to the apostolate has not yet acquired its own methodology, though the importance of it is clear. The apostolic gift is seldom innate. Few people have a natural or supernatural genius for the apostolate. But so much can be learned, and faith in the Holy Ghost does not dispense one from a humble apprenticeship in human relationship with individuals or groups. While admitting the sovereignty of grace we must also realize that it demands of us that we love God not only with our whole heart but also with our whole mind, with method, imagination and a sense of organization.

It will be the job of centres of methodology for the apostolate—still to be set up—to lay down ways and means. But one can already study past experience and distinguish the main lines.

Some Necessary Conditions for Initiation to the Apostolate

To be adequate, the initiation must be practical. Theoretical courses are not enough. There is what is called 'learning by doing' which is better than anything. All the theories one may propound about the art of swimming are not as valuable as one swimming bath. The slow, continuous apprenticeship on all fours with life remains an indispensable means of training.

Since it is to be practical, the training must be by stages spread out over a number of years.

It will produce results only if it is well adapted, made to measure, and carefully supervised. It must be integrated, as an equal partner, with other aspects of education: it cannot be relegated to the side-lines as something merely supplementary.

In the light of what has already been said, one can see that preparation for the apostolate must be widened to correspond to the complexity of the duties which it involves.

Not only teaching nuns, but also nursing nuns will tomorrow be in contact through pupils and patients with a 'hinterland' of parents, relations, ex-pupils and ex-patients, all members of the laity. This means that they will be confronted, one way or another, with the problems of the adult laity : and by corollary that they must be trained not only in their normal functions but also to act as inspiration for the laity. The scope of the training will depend on the various conditions of time and place. Depending upon whether the religious houses in question exercise their function in a Christian area or in one which has become de-christianized, there will be different problems for which they must be prepared. No stereotyped scheme can be outlined here, only a general orientation.

Apart from a real acquaintance with the problems involved, training for the apostolate involves knowing about the various organizations and movements in which today's efforts at christianization are channelled.

One must know the way the principal spiritual works are run, as well as the charitable, family and social works, and particularly the youth and Catholic Action organizations, both general and specialized, in all their modern multiplicity. If one cannot know in detail or have practical experience of everything, at least one must have some experience of those where one will have a part to play or which the bishops wish to be given special support. It is not necessary to be able to organize every type of movement, but some must be known inside-out and there must be a working knowledge of others enough to interest and direct such members of the laity as one can reach.

PROFESSIONAL TRAINING

As well as spiritual and apostolic training, professional training is also indispensable. Anyone who undertakes the role of teacher or nurse must perfectly fulfil the technical requirements. Nuns must be distinguished for their capability in this matter: it is an elementary duty in justice to those who will be confided to their care and at the same time it will indirectly increase the value of their religious influence.

On the eve of the Council, Pope John XXIII wrote to the nuns of the whole world:

> Let all those who dedicate themselves to the active life remember that it is not by prayer alone but also by works that we shall obtain a new orientation of society based on the Gospel. . . . And since in the fields of education, charity and social work one cannot make use of persons not prepared to meet the exacting conditions of present regulations, busy yourselves under obedience at studying and obtaining the diplomas necessary to overcome all obstacles. Thus, apart from your professional competence, your spirit of devotion, patience and sacrifice will be better appreciated. (2nd July, 1962.)

One must start from the idea that every nun, little gifted though she may be humanly speaking, can and must be made the best our techniques can manage. It is a duty we owe to God not to leave one single talent buried. In the developed countries means are not lacking to increase the professional training of all religious.

On the subject of teachers, Pius XII said on 13th September, 1951:

> See that you provide them with a good preparation and training corresponding to the requirements of the State. Give them generously all they need, especially books, so they can follow, even though later, the progress made in their subject and be able thus to offer the young a rich and stable harvest of knowledge. This is in conformity with the Catholic concept which accepts with gratitude all that is naturally true, beautiful and good

L

since it is the image of the divine truth, goodness and beauty.

The more capable nuns should be trained to make their voices heard in all human matters where questions are raised about the life of grace in souls and obedience to the laws of God. Abortion, divorce, birth prevention, public morals, juvenile delinquency, neglect of children, social legislation, status of women . . . nothing of all this should be unfamiliar to them. Priests of great ability have given themselves to the study of these things, but there is a dearth of nuns qualified and specializing in these fields.

The Pope excludes nothing: each nun must have the chance to put her talents to use, whether it be in teaching, care of the sick or social work. It is to be hoped that nuns have access to libraries that are kept up to date, and periodicals and current books; that they are able to attend congresses and study-groups and that these in turn are open to attendance by nuns. Study-trips in search of information should be authorized so that they can benefit from initiatives coming from outside their own neighbourhood. They should follow closely all that is being done in the various fields by government or by private enterprise.

One must not be afraid of being too ambitious. If woman today has such a place in social life, then the nun with the qualification of her professional training has a reserved seat in the same row. Wherever public opinion is formed, wherever educational laws are drafted or laws concerning the home or health, the nun has a part to play.

SOCIAL TRAINING

In order to meet what is demanded of them today, nuns must be aware of the social realities which condition the world they live in. They have to know not only the various kinds of common misery but also what causes them. They must be initiated into the great social problems which affect a man's life today. If a room is flooded, it is no good just mopping the floor: one must find out where the water is coming from and do something about it. Charity is not just an individual thing between individual persons, it is a social thing. What does loving one's neighbour as oneself mean then? It is not just a question of giving alms or being kind to individuals, it means caring about society itself and trying to even out its inequalities and let the light of justice shine.

We must not forget that one item of social progress may abolish or alleviate the material and moral misery of millions of human beings in a moment. Our nuns, therefore, must be able to situate their work in the social context of our own times, taking due notice of the psychological components of the present-day world. All this requires on their part, presence, openness, and readiness.

As teachers they will have to introduce their charges to the great problems of the working classes and all that is implied in the sombre words 'social injustice' and 'poverty'. Older girls should have close knowledge of the anguish of a family at grips with sickness, uncertainty and inadequate wages; they should have seen a slum and its promiscuity; they should know something of marginal cases which the law covers inadequately if at all; they should understand the reason for things like strikes and demonstrations. Our young people should be trained to choose their career in the light of the social service which it implies. The governing class will be powerfully influenced by the moral judgment given on the social phenomena by young adults who have to behave as Christians not only in their private lives but also in their social lives. They have their contribution to make towards the necessary redress.

Everything mentioned in the Encyclical *Mater et Magistra* should be broken down into concrete realities for our pupils. We must be able to set up social and mutual assistance teams which can teach young women of the leisured classes to seek out moral and physical distress, not with maternalism and condescension but with a deep sense of service and a clear knowledge of the requirements of social justice.

All this presupposes some training of the instructresses. It is difficult to imagine a teacher of today who is not familiar with the great currents of our age or who cannot trace the source of the profound stirrings which shake our society. Communism covers one third of the world : one cannot ignore it, and it is no good just condemning it. One must know its origin and the partial truth responsible for its birth before being in a position to condemn its atheism.

We must have the courage to face up to certain situations. As Fr. Van den Hout once wrote :

> Our western civilization has failed in its primary mission. A Christian civilization, it could and should have christianized the universe. Deeply apostate, it sought only to exploit the universe for purely worldly, even basely materialistic ends. But if the white races have been able to enrich themselves to an unbelievable degree, it has been at the price of enormous and terrible social injustices. Its cornering of wealth, its creation of an immense proletariat, its culpable lack of concern in the face of the appalling misery of millions and hundreds of millions of human beings undernourished and dying of starvation, have aroused and built up an enormous rancour that falls easy prey to the apostles of human pseudo-liberty.

The social education which is necessary for teaching nuns is also necessary in other forms for other kinds of religious who are in contact with the world. They have to accost. one by one, the souls which Providence plalces in their path and give them the best they have—the living Christ. But they must know that these souls are influenced by the world around them and do not easily resist the undertow and the many currents to which they are prey. Religious must know about these currents in order to guide their own actions and also in order to help to christianize the circles where souls are fashioned and so often de-christianized.

Mgr. Bonet, Chaplain-General of the Workers' Catholic Action in France, has not hesitated to declare:

> It is through women that working-class society will return to being a Christian society, and it is through the nun that the working-class woman will become an enlightened and effective factor in the christianization of the working-class world.

This declaration, which is valid for all social classes, is worth remembering. A social education adapted to the nature and needs of each religious congregation will give each nun an incomparable power of penetration and will be nothing but gain for the realism of her approach and the efficacy of her apostolate.

CONTINUATION EDUCATION

Continuous Education

This varied education begun during the juniorate cannot be considered as completed when the juniorate ends. Our age has started on the road of continuous education on account of the non-stop evolution of new techniques. If he is not to become very quickly out of date, an engineer or a doctor must keep abreast of developments in his field. For the doctor it is an obligation in justice to his patient who has a right to the best treatment. The acquisition of a degree or diploma is less and less the end of study: rather, it is the beginning. A French scientist made the sally that France was 'a country where one learns a great deal at school and where one acquires a number of very difficult diplomas all of which allows one to learn nothing else for the rest of one's life'. That will serve as an opportune warning against the fetishism of diplomas and against all kinds of immobilism.

Continuation studies are today equally necessary for our nuns, and Superiors must take care to see that they are carried out. If industrial concerns organize refresher courses for their technicians in order to ensure productivity, it is only right and proper that we should do the same to ensure a better apostolic yield. This is one sphere, at least, in which the children of light can learn something from the children of this world.

On the spiritual and apostolic plane this refresher education must be carried out if we are not to risk a lack of balance. Professional necessity makes a number of our religious acquire the necessary diplomas and benefit by the valuable training they afford, but if religious and spiritual education does not keep pace there is a risk that in the course of years there will be such a differential that the humanitarian aspect of their work would overwhelm the Christian aspect. What the world needs far more than Christian humanism is a thoroughly human Christianity. This means a continuous and suitable religious education. We have agreed to raise the intellectual and professional level of our nuns at the instance of the State; now the requirements of our own life must lead us to give priority to integral religious education which must be developed as far as possible.

Second Novitiate

A certain number of congregations arrange in some form or another for a sort of second novitiate. This idea should be expanded and deepened. Straightway one comes upon the necessity of grouping together the smaller congregations so that they can have the benefit of these refresher courses which require a directing staff difficult to find on the spot. Such courses would be taken regularly at intervals still to be determined and should cover every aspect of the life of our nuns, that is to say not only their spiritual life but also their apostolic, professional and social life. Each aspect has its own requirements. Deepening the spiritual life can be done in a special house, in silence and recollection—in retreat, in fact. But other forms of refresher training could not be carried out in those conditions. In fact it will be necessary to learn new apostolic techniques and bring old methods up to date, in the same way as nuns will have to review and bring up to date their theoretical and practical knowledge of their professional and social work. All this involves contacts, movement—in fact courses. What will be needed will have to be worked out for each type of refresher course. The four aspects—spiritual, apostolic, professional and social—must all be dealt with, but not necessarily all within one year. The incidence and programming will have to be studied.

This will demand sacrifices, adaptations and an imaginative effort in order to make nuns, not always easily replaced in their jobs, available; but the results will be worth the trouble involved.

Nuns have dedicated their lives to God in their communities; it is right and proper that the communities do what they can to foster a renewal which will increase the yield both natural and supernatural of these dedicated souls. Steps will have to be taken to ensure that all nuns are reached. The Superior and her assistants must love each member of the community individually and develop all her energies and all her natural and supernatural gifts. All this is part of the permanent education towards which we must strive. There is nothing like it to dispel the danger of spiritual sclerosis and to ensure that God is served with joy.

"A soul which lifts itself up lifts up the world." A nun of whom the very best has been made in all these four aspects will draw after her an ever growing number of souls. Superiors have a duty to be ambitious for their daughters. They were chosen by God and God has in reserve a super-abundance of grace to allow them to respond adequately to their vocation.

12

HOW TO HASTEN THE
RENEWAL FROM WITHIN

THE renewal proposed in these pages depends on the coming together of the efforts made within and without, that is to say the efforts made *by* the congregations and *for* the congregations.

In this chapter we should like to analyse what can be done within the organizations to speed and promote advancement of the religious and, as a consequence, allow her to benefit from the pastoral revival which the Church hopes to introduce at all levels.

This advancement depends on the attitude which will be adopted on the questions of enclosure and allocation of time. Unwillingness to move in these two fields guarantees in advance the failure of any attempt at renewal. On the positive side, integration of the apostolic meetings we shall describe into the life of a community will be a firm guarantee of the will to adapt.

As we said before, the necessary revisions must not be the work of individual nuns—their vow of obedience requires them to obey a Rule, not to change it; but it is legitimate for them to collaborate in bringing about changes once competent authority has agreed in principle. The competent authority is alone responsible for directing the revision process. To do so, it is necessary to enter into all the aspects of renewal. What is, in fact, the best way to set

about revision? We believe that the initiative lies in the
first place with the General Chapters.

GENERAL CHAPTERS

The General Chapter of each congregation provides for
the periodical revision of the constitutions. The instrument
is available, therefore; let us see how it works.

A congregation is governed as to ordinary affairs by one
of its members, the Superior General, assisted by a Council,
and as to extraordinary affairs by the General Chapter.
Necessary revisions are first of all the concern of the
General Chapter—the highest authority of the congregation.
Part of its normal function is to see that revisions necessi-
tated by new needs or new circumstances are carried out.

Previously, I was told by the Superior General of a very
important congregation, it was almost exclusively a question of maintaining tradition. It is only just under Pius XII
and principally since the first Congress of Religious (1950)
and that of Superiors General (1952) that anything has seri-
ously been attempted in the way of adaptation. I asked
him to tell me what were in his eyes the most important
reefs to be avoided at a General Chapter, and he gave me
the following points.

Chapters are sometimes improvised and almost always
insufficiently prepared. Improvisation occurs when the
Chapter is called only on the death of a Superior General.
Lack of preparation is due to lack of continuity of information between Chapters and a lack of information
necessary for the preparation of the Chapter. Adequate
consultation between members is needed, as well as preliminary studies and enquiries. Also, too long intervals
between Chapters are undesirable in our times when
everything evolves so rapidly. Moreover, the Chapter
should be really representative.

Before the Chapter

Representation is often one-sided. This arises from the preponderance of *ex officio* members over elected members, of Superiors over ordinary members. A certain concept of religious obedience and of what is 'suitable' leads a number of religious to send Superiors as delegates to the Chapter rather than choose them principally for their competence, to choose older rather than younger men, and in the case of congregations working in several fields, to overstress one or other of them.

Another reef to be avoided is the tendency to attach no importance to anything except the elections and to treat other business, i.e. problems of adaptation, etc., as subsidiary; or again to put all the emphasis on religious problems at the expense of apostolic questions, or to be concerned only with maintaining tradition, not with adapting it.

During the Chapter

Once the Chapter is in session there are other dangers to be avoided. The Superior General and Council must be careful to remember that, in Chapter, they have no power of decision. The delegates to the Chapter are in fact there to check their government and to ensure that the interests of the Community prevail. There is danger that the Chapter will fail in its object due to a mistaken sense of deference among its members who dare not bring deficiencies to light or make suggestions for fear of upsetting some Superior, or of seeming to be too revolutionary. Members must not lose sight of the fact that they are there not to listen to the Superior General dictating but to deliberate and legislate in common. The temptation to passivity or easygoingness must be resisted. In order not to make the Chapter last too long, one is inclined to refer problems 'to the wisdom of the

Superior General and Council' or else to rush through them at high speed.

Preparatory Studies

The above comments seem to us of the greatest importance. To emphasize one of the points brought out—insufficient preparation—one might use as one's inspiration the preparatory work for the Council. The pre-preparatory committees followed by the preparatory committees cleared the ground and analysed the 9,000-odd suggestions received. Similar preparatory work for a General Chapter would have to be carried out with the greatest freedom, discretion and frankness. To get at people's real reactions is not always easy, for people do not like to say, to their faces at least, things which they think may offend their Superiors. Preliminary studies will have to be made of very precise questions, with open eyes and ears. It may be advisable to question certain reliable lay witnesses. The sort of question to be put might be, 'What is it in what she sees of religious life that puts a modern girl off and spoils her vocation?'

The question does not cover everything, of course, and is not intended to evoke all the multiple causes which, from the outside, militate against vocations, but it is direct and precise and is part of the main question how to hasten renewal from within.

It is easy to see how much more realistic and effective a General Chapter would be if it were preceded by a thorough enquiry addressed to all members of the congregation.

REPLACEMENT OF SUPERIORS

If we are to hasten the renewal from within, a point of primary importance is that of the duration of the term of office of Superiors. The Central Council of another great congregation whom I consulted on this point, sent me a note which summarizes considerable experience in the matter and which seems pretty well in harmony with what we have already cited about General Chapters.

It is noted that with few exceptions most female congregations have a Central Council composed of a majority and sometimes a totality of older members. Commonly enough elections for renewal of terms of office have their own particular psychology : a nun very easily feels that she must not be ungrateful to a retiring Superior even if there is an obviously competent candidate to take over her office, and so she votes for the present holder. Another point is that although freedom to vote as one will is theoretically guaranteed there is often a fear, founded or unfounded, that in the end the votes do not remain secret, and this obviously inhibits freedom. Finally, most of the time a serious renewal is almost impossible owing to the fact that the Superior General, once in office, nominates at her discretion all the general staff of local Superiors who are therefore devoted to her and who are *ex-officio* members of the next General Chapter.

Present legislation is in itself and quite independently of personalities (who are not called in question) an element of immobilism and a major obstacle to renewal from within.

It is to be hoped that any revision of the Code will take this situation into account and that remedies for it will be studied. It will be enough to revise a few points.

Election procedure should be improved so as to guarantee freedom and secrecy.

Capitulantes, i.e. members of the Chapter called upon to elect the new Superior General, should comprise at least as many elected delegates as *ex-officio* members.

There should be an increase in the percentage of votes

required to re-elect an office holder and there should be, in accordance with the wishes of the Sacred Congregation for the Affairs of Religious, no recourse to the process of postulation which allows terms of office to be prolonged by special permission beyond what the rules provide.

Juridical note should be taken of the Church's expressed preference for periodic changes of government, an age limit should be fixed, and the wish should be expressed that as a matter of principle the retiring Superior should not, as is the present tradition, be a member of the new Central Council.

Legislation forbidding direct or indirect influence on elections should be strengthened and be stricter in allowing a retiring Superior to be nominated as Superior to another house which has the result of a person's being a Superior for life. Nuns find difficulty in accepting that a one-time Superior is no longer such and must take her place in the ranks.

All this has been on the juridical plane. It goes without saying, however, that any renewal must be supported on the psychological plane by the creation of a mentality which will welcome these new ideas. One may hope that nuns, being clearly aware of the Church's wishes today, would be happy to adhere to directives which they know to be inspired by the greatest good of their communities.

H.E. Cardinal Larraona, at that time Secretary of the Sacred Congregation for the Affairs of Religious, expressed himself with all necessary clarity on this subject in 1952, addressing the Congress of Mothers General in Rome:

The Sacred Congregation is not in favour of re-elections beyond the term stipulated in the Constitutions: more, it is in principle opposed to them. Superiors and *Capitulantes* are obliged to observe the law of the Church just as their subjects are. The prolonged retention of one person in an office tends to prevent the training of other Superiors and restricts the choice too narrowly. . . . In the case of a possible 'postulation' . . . the judgment of the Sacred Congregation will furthermore be strict, since the confirmation of a re-election beyond the fixed term will constitute an exception which should be rare.[1]

Superiors ought to take pleasure in preparing for their relief, in distributing responsibility, in regularly introducing new blood into the government in order to be in better tune with the times—God knows, we are evolving rapidly enough in these days of space flights!

The Mothers General during their Congress in Rome in 1952 expressed the following wish:

> For the continuous training and perfecting of Superiors . . . one must not exclude the younger nuns from office, must not make conditions not made by Canon Law, not persist in re-electing the same persons, for the Church's mind is that the law and constitutions of the congregation should be obeyed and they find it good that there should be alternation in the office of Superior so that Superiors shall not be deprived of the benefits of obedience. Let it be remembered that where other conditions are equal or nearly equal between a Superior in office and a new candidate, it is preferable to elect the latter. Regrettable crises will thus be avoided and one will have a greater number of religious trained in governing.[2]

What a good example, incidentally, can be shown by a retiring Superior who receives an ordinary obedience and shows by her joyful humility the truth of what she has been teaching. If she is later again in office, she will be all

[1] Sacra Congregatio de Religiosis, *Acta et Documenta Congressus Internationalis Superiorissarum Generalium.* 1952, p. 271.
[2] Ibid., p. 299.

the better qualified for having once more seen things from the other side and will be nearer to her subjects and better able to serve them.

PERIODICAL REVISION

Periodical revision is normal and healthy; a rule must be adapted to reality—which means that it must be brought up to date to keep pace with the developments of reality.

In fact it will not be the Constitutions so much as the Directories and Customaries that will be subject to revision. These are the *vade-mecum* of everyday life: the spirit of the Rule is brought to life in a thousand concrete details. It is they which often either inhibit or encourage liberty of apostolic action. They must be carefully checked, for their provisions control the whole life of a congregation. It often happens that with the passage of time they become an obstacle and a hindrance to enterprise instead of a stimulus and support.

Once Customs have been in use for ten years, one may be sure that danger level has been reached and that some paragraphs at least have no longer any connection with reality.

In the course of the necessary revision a negative work is first necessary—the elimination of everything that hinders the apostolic developments previously described. After that comes the task of making positive provision for the four-dimensional training we have spoken of.

APOSTOLIC MEETINGS

The integration of apostolic activities must have its official place in the community life. Such activities cannot be left to chance nor treated as poor relations. They cannot be left to amateur efforts, much less be not subject to obedience. On the contrary, it is necessary that this sort of activity be incorporated in the normal life of the community and not just be a fringe activity. The hours set aside for it must be part of the normal routine and take their place among other duties as spiritual exercises of a different outward form but the same animating spirit. It is the whole community that is involved in it so it will be reasonable to make room for apostolic meetings which will occur periodically, be properly regulated and have a 'staff'. These meetings, for which provision will be made in the time-table, will consist of a 'sharing' of apostolic activities with mutual exchange of information and criticism and co-ordination in the distribution of tasks. It is easy to see why the community should meet at regular intervals to consider as a body the way in which each individual effort should be directed and to follow up and co-ordinate results. Without something like this there would be the tendency for the work to get no further than the expression of personal inclinations and unrelated impulses. A community will never really get going in this new dimension of the apostolate unless each of its members is assigned a definite task which will cause her to discover new possibilities of religious activity and of which she will be called upon to give account. The meetings are essential for cohesion of the work and for the spirit underlying it. They could be compared to spiritual exercises, a time for prayer which is fixed but in no way excludes prayer at other times: on the contrary, one prays at fixed times in order to accustom oneself to the climate of prayer. Similarly, the apostolic meetings we are talking about are a powerful means of placing the soul in a state of constant apostolic awareness.

In the case of a large community it would be better to divide it up into groups or 'apostolic teams' of ten to twelve nuns. This would avoid over-long meetings, would increase the yield of each member and give her the maximum of personal responsibility. To avoid dispersion of effort, there could be periodical meetings of all the teams or their repre-

M

sentatives. All this, of course, must come under obedience:
it will be the Superior's responsibility to nominate the
members and officers of each group and to exercise overall
control of their activities.

The teams will give the community all the advantages on
the human plane of group psychology. The necessary discretion will allow individual nuns to put the best of themselves at the disposal of the common good. The meetings
will be a stimulus for community life. Once they have been
successfully tried out, it will be up to a General Chapter to
give them official status.

It would be interesting to hear about enterprises of this
nature so that others can benefit from the experience.[1]

APOSTOLIC GUIDANCE

So far we have dealt with getting the renewal under way.
Once the principle is accepted, it may be useful to have
qualified guides or advisers who can both start enterprises
and assist in integrating the apostolic meetings into the
community life. This sort of task could be given to 'apostolic' visitors, that is to specially trained nuns who would
be able to place their experience at the disposal of communities. They could be found within the communities or
'borrowed' from communities which have the necessary experience. Their assistance would greatly facilitate the setting up of new enterprises and the avoidance of mistakes.
A great many problems are much more easily dealt with
by someone on the spot and a 'visitor' would be a great
encouragement to face up to all the 'impossibilities', which
usually disappear if one has the courage to believe that
they will.

---

[1] Information and details about the integration of apostolic meetings into community life will gladly be furnished by the Conseil
Pastoral Diocésain des Religieuses, l'Archevêché, Malines.

13

HOW TO HASTEN THE
RENEWAL FROM WITHOUT

IN order to make the most of her own apostolic possibilities, the religious of today needs help from outside and she should appeal for any possible collaboration as she appeals to the authorities on whom she depends.

An exceptional opportunity of making a decisive turning point in the development of the apostolic aspect of the religious life is available. It is none other than the Second Vatican Council now being held in Rome with the object of encouraging a spiritual renewal throughout the Church.

THE COUNCIL

The bishops have expressed or passed on several thousands of wishes. It is to be feared that the Superiors of our congregations have not made their voices heard enough nor made known their wishes on the subject of the things that hinder or inhibit their vitality and power to radiate Christ. But it is not too late: what was not done before the Council can be done afterwards. It would be reasonable to suppose that a place will be found for them on some post-conciliar consultative committee and that the thousand complex questions to be solved about the religious life of women will not be left in exclusively male hands. In any case a possibility of dialogue is desirable.

But that is for the future.

But right now the Council could cause a decisive step forward to be taken in the matter of the apostolic advancement of nuns by deciding to appoint a committee to study the problem. The Council could determine some basic guiding principles and instruct the Committee for the revision of the Code of Canon Law (which will sit immediately after the Council) accordingly. It is usual for the Code to be amended from time to time, for the law should be patterned on reality and not the other way about. There is normally some time-lag between the need for revision and its being made. A Roman canon lawyer used to say by way of encouraging the promoters of useful changes, "You must be the torrent: we lawyers can only provide the bed for you". Put in another way, the canon lawyers cannot do anything useful unless they know what is wanted and are encouraged to do the necessary work. In the hope of achieving this, it could be respectfully but insistently demanded of the Conciliar Fathers that they :

Affirm the principle of the new role of the nun as animator of the feminine laity both young and grown up.

Express the wish that adequate theoretical and practical training for this role be given to young nuns.

Express the wish that nothing in the Constitutions should be a hindrance to the necessities of the apostolate in the world of today and that Constitutions be revised in this sense.

Express the wish to see the religious life so organized that this new function becomes an integral part of community life and has by right its recognized place therein through the organization of special meetings to control and foster specific activities.

The Council's agreement to this would be of incalculable importance and would furnish a decisive impetus.

NATIONAL AND DIOCESAN FEDERATIONS

Union of Major Superiors

Adaptations within communities could also be greatly helped from outside by already existing organizations for collaboration. Diocesan federations of religious are, like the national federations, the obvious instruments of renewal. So also are the Unions of Major Superiors, which, at the instigation of the Holy See, have come into being in many places. Their task is to overcome inertia, routine and the spiritual or intellectual laziness which everywhere favour a policy of no change. They permit religious to compare problems, exchange views and to work out concrete proposals to be submitted to higher authority. These periodical meetings and exchanges provide a remarkable opportunity for progress. They were in fact started by the Holy See with an eye on the changes that will be necessary. No doubt a part of the problems to be studied will deal with the internal arrangements of the religious life, but most of these problems affect the apostolate. They must be studied with the constant idea of being always one step ahead of apostolic needs while still respecting the principal object of each institution.

One cannot fail to welcome a flourishing set of unions grouping together religious engaging in school and hospital work and parish schooling. An enterprise has been started which aroused justifiable hope. Books and periodicals, congresses and study-sessions are attacking the problems of the religious life with courage and objectivity.

The Holy See is generous in its encouragement. The question of the nun's position has been brought out into full daylight. One must hope for active participation on the part of qualified religious in the search for concrete ways and means of solution. Listening to lectures is not enough. Investigations must be carried out and practical proposals put forward; it may be a long business but it is a work that should command the devotion of all.

Particularly welcome is the institution of training sessions for Superiors.

Training Sessions for Superiors

Behind every serious reform lies the problem of training Superiors. Until recent years, election to office was all there was to it. People are beginning to realize that there is something missing and arrangements for training Superiors are becoming more numerous. This is a step in the right direction, particularly since the training is not only for Superiors as such, but also for assistants, mistresses of novices, all those who in one way or another have the fate of the community in their hands. The Superior must be able to make a team out of her Council and must show by example what it means to take over collective charge of a community. Several heads are better than one—not only for intelligence but also for wisdom, and above all to harmonize the various complementary aspects of community life and to avoid the stifling effect of maternalism centred in one person. Authority grows in stature by giving due importance to its subordinates. In any case the Church in her laws has always emphasized the part to be played by Councils in the government of religious houses. Any training which will help Superiors to work as a team with their Councils will be of the greatest benefit.

Training Centres

The renewal can also be greatly helped by the establishment of training centres for nuns. Some centres have already been established to provide suitably prepared nuns with the opportunity of acquiring a solid philosophical and theological culture. Why should it not be possible for nuns who have taken this instruction to run retreats or days of recollection, or direct spiritual and apostolic sessions of lay-women or girls? *Regina Mundi*, the Pontifical Institute for nuns in Rome, where teaching is in four languages, indicates by the very fact of its existence what the mind of the Holy See is in this matter. The time will come no doubt when a complementary practical introduction to apostolic work will complete the object of this Institute.

It is desirable that centres of *apostolic methodology* should also be opened for the teaching in theory and in practice anything that can increase the power of penetration of the apostolate.

One must in fact learn how to carry out the apostolate and how to teach others to do so; which means knowing how to find apostles, how to convince them, and how to train them. This in turn means knowing how to organize and co-ordinate an apostolate on many fronts. This can only be done by attending courses in order to learn, from the inside, all about the main apostolic movements which the hierarchy considers to be necessary in any given area.

A great deal of energy has been given to studying theoretical problems: the same sort of effort is now needed in the practical sphere to find concrete ways and means. It is not enough to know a message and understand it and formulate it to oneself, one must learn by practice—for it is something to be learned—how to transmit the message naturally and supernaturally to others. Schools of this sort, either independent or attached to some existing institutions, would be of great help in the training of nuns. They should be for the apostolate what the *Lumen Vitae* centres are for catechesis. But, let us say it again, they must be based on *practical* training; the temptation to be content with lectures and theoretical instruction must be resisted.

Naturally enough one thinks about the part that could be played by teacher training schools (*Ecoles Normales*) where tomorrow's teachers are trained. If it were possible to prepare a course of apostolic methodology there and grant a certificate, one would have established something by way of a cadre training which would be an important step towards increasing the apostolic yield in our schools.

One would like to see schemes started to establish these centres of methodology and to study their syllabus and its application. Experience would show the best methods for this new sort of instruction rendered necessary today by the vast apostolic problems facing us and, in the case of religious, by the part they have to play as animators of the laity around them.

Regrouping

The regrouping of dispersed forces is a potent means to increase their efficacy: dispersion and the frittering of efforts are the two main causes of weakening of effort. The need for regrouping communities that are too small on their own is admitted: the Holy See is favourable to it as a matter of principle and makes provision for many different forms of it. One of these is federation, which has been introduced for enclosed nuns and which allows complete autonomy but makes provision for certain common services. Besides federations there are unions, which unite congregations of the same type. There is a whole range of possibilities combining autonomy and unity in various proportions.

There is also room for complete fusion either by two small congregations uniting to form one larger one, or by the absorption of a small congregation into a larger one. There is no doubt that this sort of regrouping can appreciably speed up the desired reforms and permit adequate training. Each case is a special case and there is no question of studying all aspects of the matters here: all we wish to do is to note the place of regrouping in the general perspective of renewal: it is by no means a negligible place and the idea merits favourable consideration and careful study by all congregations with reduced numbers.

It is easy to see the benefits to be expected from such fusion; greater choice of personnel to fill fewer jobs, one person per job and one job per person, freeing of forces and the disposal of property, real and otherwise, which may be used to extend the field of action.

A real love of souls combined with genuine detachment from self will lead the Superior General and her Council to forget themselves and their natural and understandable predilections and base their decisions on the needs of the faith and the interests of God.

One must not wait passively until higher authority takes the initiative nor till things have gone so far that fusion is no longer possible. Let all Superiors General to whom this applies have the courage to take the first step, to set out the markers, and to plan for a union which would be for the good of all. Once they have created the right atmosphere and made the first approaches, the work of the ecclesiastical authorities is greatly simplified—they have only to sanction a union which has already taken place in spirit.

In cases where the grouping together of small communities is found to be impossible, it is desirable, as we have said, that they join forces to establish a novitiate common to several congregations (which will remain independent) and to arrange for the second novitiate about which we have spoken; our congregations have everything to gain by getting away from isolationism and particularism. The habit is happily becoming commoner of arranging retreats and recollections for members of several congregations together. Experience shows that exchanges between congregations are a great stimulus and source of fruitfulness.

Regrouping is the practical way of allowing nuns to get the maximum benefit from good lecturers and preachers who are too few in relation to the number of religious communities to be able to meet all the demands on them as things are at present.

PROMISES FOR THE FUTURE

It remains for us at the end of these pages to look to the future and to see what the apostolic advancement of nuns holds of hope for re-christianization.

We can say that tomorrow's society will be what our Christian families are, and that they will be the image of the nun who has trained her who will be tomorrow's wife and mother: a Christian woman conscious of her apostolic duties.

An immense responsibility but also a vocation all the more magnificent because its place is in the very heart of natural and supernatural life.

In the Family

"Society will be reformed by woman, or it will not be reformed at all," wrote Fr. Rijckmans.[1] A phrase of wide significance.

It can be applied literally to Christian society, which will be what the Christian woman of tomorrow makes it, for it is she who builds, stone by stone, both the earthly city and the city of God.

Archetype of the social cell, the family is also the first cell of the Church. It is the first sanctuary, built of living stones, where God establishes a dwelling.

To build a family is to build the Church, to work, basically, for the salvation of the world.

The nun who has understood these pages will always see the family as the final objective of her activities. She will

[1] *L'aide des Laïcs au soin des âmes*, p. 90.

instinctively place the pupils she teaches or the sick confided to her care in their family context. Every nun being essentially an educator—whether she belongs to a teaching or nursing order—she will carry her teaching to its end, that is up to and including the family, whether it is a question of creating a Christian home, consolidating it or bringing it back to health. As an educator, she will carry out this function to its full extent in relation to lay-women whose full capacity for radiating Christianity she will bring out.

Usually a nun comes from a thoroughly Christian home. She should keep the picture of this home constantly before her eyes as an incentive to make Christ live in their homes. She must not regard the picture as a relic of the past but as something to be projected into the future in the service of others.

What a wonderful thing before God and man is a family which fully accepts the Kingship of Christ.

They pray together and their common prayer links souls together by ties stronger than blood.

They kneel side by side at Communion and allow Our Lord to transform each one of them into His own image as they nourish themselves on Him.

They work together in the same spirit of duty. The children, seeing Christianity lived by their parents in every aspect of life, acquire without realizing it the spirit of the apostolate and of devotion which is at the root of all generous vocations.

The Source of Vocations

This picture of a Christian family, albeit incomplete, is kept by the religious in her heart so that she may give it to the world. Through the children she teaches or the sick she cares for she is working within the family framework, even though indirectly. Children, the sick, the aged, all open a door for her to their families: she can never forget it and she must exploit it to the full.

By her contribution to the christianization of families the religious is preparing the ground in which great vocations can germinate and mature, the good ground of the parable which receives the seed and yields a hundredfold harvest.

These will be vocations to the lay apostolate, to entering religion, and to the priesthood.

Lay Vocations

Lay vocations on the one hand prepare for, support and extend the action of the priest, and on the other christianize their own circles by the natural and supernatural power of radiating Christ.

Who can be blind to the immense field of action open to nuns trained for this work and called upon to arouse, organize and sustain lay-women who (let us say it again) without them would never be mobilized on a scale to meet the requirements.

Religious Vocations

Our Lord will not be outdone in generosity. He will respond to the generosity of active congregations by assuring them of the new blood the need for which is now so urgently felt. They will be the first to benefit from the increase in vocations. The sight of their apostolic courage will be an inspiration and a great attraction for the young.

This is also valid for contemplative vocations which are vital to the Church to supply the high tension current feeding all apostolic efforts. Contemplatives have chosen silence and real separation from the world for love of God and for the salvation of the world. Active religious have chosen speech and the direct transmission of God, but the impulse of love is the same. The silence of one gives power to the words of the other; those on their knees give strength to those who must march. There are reckoned to be some 60,000 contemplatives in the world. In relation to the total of 1,000,000 or so women in religion altogether they can be regarded, as one writer said, as the 6 per cent. return for capital which God reserves to Himself.

Priestly Vocations

Finally, it is true for priestly vocations, the highest expression of God's love. Without the priesthood Christ's visible mediation would disappear from among us, and contemplative vocations would die of inanition, sacramental life would wither. Without the priesthood there would be no Eucharist. Without the priesthood humanity would die of spiritual starvation.

All these graces come from God of course, but also through the co-operation of man. It is the home that gives man's first answer to God.

When she consecrates her life to the young and the sick and through them to the salvation of the family, the religious places herself at the very heart of the religious and moral future of a country.

She holds the keys of the Kingdom. The Church can say to her in the words of Scripture, "My fate is in your hands."

The Keynote is Joy

In agreeing to carry out her teaching role to its proper talking about, the religious not only does an invaluable service to the world, she will also find the secret of a more complete religious development.

In agreeing to carry out her teaching role to its proper conclusion she will have to exercise to a greater extent the theological virtues of faith, hope and charity. Nothing will help her spiritual development more. The heart can only expand when the horizon widens and one can breathe

deeply of a tonic air. A great joy will follow her acceptance, and a great hope will arise in the Church. Let her not
be afraid!

When St. Peter was told to walk to Jesus on the waters,
all went well so long as his eyes were fixed on those of the
Master. It was only when he turned his eyes away to
estimate the changes of the wind that he became afraid and
began to sink. And Our Lord said, "Why didst thou hesitate,
man of little faith?"

This passage of the Gospel has a special meaning for the
religious authorities called upon to promote the apostolic
zeal of our active congregations. The Master speaks to them
again today through those who speak in His name, "Sail
across the water, pay no attention to contrary winds or the
movements of the waves".

Our Lord will be at hand to steady their steps and to
overcome the 'impossibilities' of mere human wisdom. He
will be there to encourage them with His smile, with His
fortifying grace, with His power which laughs at the world
and Satan. He will be there to give their hands the strength
of the Resurrection itself to the extent that their faith is
alive and creative.

The holy women who approached the Saviour's tomb on
Easter morning asked themselves how they were going to
roll away the stone. But they started out without having
found an answer to the problem, and the Lord solved it
magnificently for them.

We have the right and the duty to count upon God's
grace since He does not ask the impossible and gives what
He asks for. The essential is to start out, with faith.

It is said of Blériot that on his first flight he placed in
front of him in the machine a statue of Our Lady inscribed
'*Regarde et prends ton vol*' : 'Look about you, and then take
wing.'

This is the message with which we wish to end.

Let our nuns, too, fix their regard on the Blessed Virgin. Let them accept courageously the necessary changes and sacrifices. Let them offer themselves to continue in the contemporary world the spiritual motherhood of our Lady under whose aegis the Vatican Council has so auspiciously begun its work of pastoral renewal in the Church of God.